"There is one thing stronger than all the armies in the world, and that is an idea whose time has come."
—Victor Hugo

PRAISE FOR *IF YOUR MOUTH COULD TALK*

"If you have a mouth, this book is for you."
—Paul Krebsbach, DDS, PhD, Dean and professor at UCLA School of Dentistry

"Being a quadruple board-certified physician practicing internal medicine, pulmonary, critical care, and sleep medicine in a university setting feeds my desire and dedication to constant learning by teaching others. I feel this similar passion for education when reading *If Your Mouth Could Talk.* Dr. Kami Hoss has a fun and captivating way of connecting oral health with the entire body. He prioritizes sleep as a key foundation for good health and wellness, which is a mirror image of what I believe in and practice daily. He does a great job in telling a story of how oral health and sleep are more closely related than one might suspect. This book was a delightful guide to the importance of integrating oral health with the entire body with a nice focus on sleep medicine. Dr. Kami Hoss takes the reader on an informative and enjoyable journey into the world of oral health. Being a proud father of 3 kids who are not always fans of brushing their teeth, Dr. Kami Hoss has expertly written a book so parents can proactively improve the health, happiness, and future of their children."
—Raj Dasgupta MD, FACP, FCCP, FAASM, associate professor of clinical medicine, pulmonary, critical care, and sleep medicine

"As someone who promotes health and fitness for a living, I'm blown away by just how little I knew about the depth of the correlation between dental health and overall health before being introduced to *If Your Mouth Could Talk.* Getting my daughter (and other children) to eat vegetables while limiting sugar is easy and even fun for me ... getting her to brush her teeth is not. So, I am extremely grateful to have found Dr. Kami Hoss to make sure parents understand the importance of creating whatever fun it takes to prioritize oral health and to show us how to get the right dental (aka life) start!"
—Devin Alexander, *New York Times* bestselling author and celebrity chef

"As a biochemist and physiologist, I've come to realize the most important health care practitioners in our society today are not physicians but rather dentists. So much of the chronic health conditions that plague Americans today can be explained by what goes on in the mouth. Anatomical issues can affect our breathing and sleep, oral infections and oral dysbiosis can affect our cardiovascular and nervous systems. Very few people understand this connection. Dr. Kami Hoss has assimilated this life changing information into an easy-to-read book that should be taught in every major health care curriculum. Furthermore, it should be required reading for anyone interested in taking control of their own healthcare."

—**Nathan S. Bryan, PhD, author, inventor, and entrepreneur**

"As a family medicine physician caring for both kids and adults, and as a parent myself, I found Dr. Hoss' book informative and necessary. For the past decade or so, the medical field has begun to discover and accumulate numerous studies confirming the link between oral health and chronic disease. Dr. Hoss has gathered all of this data, consolidated the information into one comprehensive resource, and translated the knowledge in a relatable way for both parents to understand and physicians to apply.

It is a particularly valuable resource for concerned parents with persistent, unanswered questions regarding the health of their children, most especially children with sleep and behavioral disorders, and those who suffer from a chronic illness. However, the information Dr. Hoss presents is relevant and useful to all of us, regardless of age and parental status. This may very well be some of the most important information you will ever learn."

—**Sanaz Majd, MD, family medicine**

"The teaching of oral development and dental care is lightly emphasized in medical education. Dr. Hoss has written an easily accessible and thorough treatise which fills this void. Medical professionals and the general public will benefit greatly from reading his wondlerful book."

—**Adam Breslow MD, MBA, pediatrician**

"Thank you for the wonderful inspiration."

—**Victoria Tate, actress and producer**

"As a doctor myself, as well as a patient who has lived 59 years, one trait that differentiates the great doctors from the others is a passion for one's work. Not only does Dr. Kami have and display deep passion for his specialty, but he understands the human body and how his area of expertise fits in with the overall health of the individual. Beyond his passion and understanding, Dr. Kami possesses other essential traits of a great doctor that are evident as soon as you meet him: intelligence, respectfulness, kindness, and empathy/sympathy."

—**Dr. Rand McClain, D.O., sports, rejuvenative, regenerative**
("Anti-Aging"), cosmetic, and family medicine

"I learned so much from reading *If Your Mouth Could Talk*. It has raised my awareness of the huge impact of oral health on general health. The entertaining prose and excellent explanations made it an absorbing and fun read. I highly recommend it to parents and anyone interested in learning more about oral health!"

—Elizabeth L. Yu, MD, pediatric gastroenterologist

"As a medical doctor and functional medicine expert, I have devoted years to the relentless pursuit of learning how to optimize the body's innate healing ability. It is encouraging to see the same passion and desire in Dr. Kami Hoss. His book, *If Your Mouth Could Talk*, does a phenomenal job of connecting the dots between oral health and whole-body healing, enabling us to better understand the connection between the oral microbiome and how profoundly it impacts our overall health. Dr. Hoss' ability to relay this well-researched critical information, in both an engaging and insightful way, is commendable. I highly recommend you pick up your copy today!"

—Jill C. Carnahan, MD, ABIHM, ABoIM, IFMCP

"The most important book you'll ever read!"

—Felix Solis, actor, director, and producer

"Dr. Kami Hoss has done a masterful job of distilling the science, while giving easy to understand details of why the way you breathe is so important to the sleeping process. I love his "Tongue Toning exercises" and I am making them part of all of my patients' rituals for better sleep. In Chapter 3 he has given one of the best explanations of Sleep Disordered Breathing and Upper Airway resistance I have ever read. This is a book I will have in my office and refer to frequently."

—Dr. Michael Breus, PhD, ABSM, FAASM, double board-certified clinical psychologist and sleep specialist, and founder of thesleepdoctor.com

"Every child should have parents who read this book!"

—Alicia Coppola, actress

"Dr. Kami Hoss writes a brilliant depiction of the connection between our dental health and our overall health and wellness in *If Your Mouth Could Talk*. As a therapist, I constantly stress the vital importance of implementing self-care on our daily routine and it starts with taking care of our bodies and our oral hygiene! It's a must read for every parent that wants to role model a healthy lifestyle for their children and have them follow in their footsteps."

—Dr. Kim Van Dusen, LMFT, RPT, CEO/owner/founder of The Parentologist

"I am truly astounded by the amount of medical information Dr. Hoss gathered and synthesized in this book. It is well-moored in history and the latest science and so engaging to read, I couldn't put it down. Dr. Hoss follows the development of oral disease from the beginning of humans on Earth until the 21st century. Everyone—from a lay person to a medical professional—will benefit from *If Your Mouth Could Talk*. I have been a physician since 1978. Today, I work in a level 3 neonatal ICU, and I learned so much from Dr. Hoss, especially from the section on breastfeeding.

I highly recommend this book. It needs to be part of the curriculum for every medical and dental training program. I would like to thank Dr. Hoss for developing such a valuable resource for us all."

—Hamid R. Movahhedian, MD, FAAP, Chief of Neonatology, NICU Medical
Director, and chair of the Department of Pediatrics at Tricity Medical Center

IF
YOUR
MOUTH
COULD
TALK

IF YOUR MOUTH COULD TALK

*An In-Depth Guide to Oral Health and
Its Impact on Your Entire Life*

DR. KAMI HOSS

BenBella Books, Inc.
Dallas, TX

BENBELLA
BenBella Books, Inc.
10440 N. Central Expressway
Suite 800
Dallas, TX 75231
benbellabooks.com
Send feedback to feedback@benbellabooks.com

BenBella is a federally registered trademark.

Printed in the United States of America
10 9 8 7 6 5 4 3 2

Library of Congress Control Number: 2021049907
ISBN 9781637740361 (hardcover)
ISBN 9781637740378 (electronic)

Editing by Trish Sebben Malone
Copyediting by Michael Fedison
Proofreading by Doug Johnson and Cape Cod Compositors, Inc.
Indexing by WordCo Indexing Services
Text design and composition by Aaron Edmiston
Cover design by Kara Klontz
Cover image © Shutterstock / Stockobaza
Printed by Lake Book Manufacturing

To Dr. Nazli Keri, my wife and partner in work and in life.

And to our amazing son, Aiden.

Thank you for inspiring me to write this book and for your love, patience, and support throughout this long journey.

CONTENTS

A NOTE FROM
THE AUTHOR

This book is unlike any book you've read before. It has no gimmicks, no silver bullets, and no outrageous claims. Instead, what you'll find in the following pages is a wealth of well-researched, scientifically backed information that will empower you to improve and extend your life. It will shock you, open your eyes, and inspire you to take better care of yourself. Most importantly, it will give you the tools with which to raise healthier, happier, more successful children.

My name is Dr. Kami Hoss. I'm an orthodontist and dentofacial orthopedist, specializing in diagnosing, preventing, intercepting, and treating dental, jaw, and facial irregularities. My wife, Dr. Nazli Keri, is an incredible pediatric dentist, and together, we founded our practice: The Super Dentists™—one of the leading pediatric dentistry, orthodontics, and parent dentistry practices in the country. For over 20 years, we've been treating patients at The Super Dentists and we've witnessed the profound impact oral health can have on a person's life and future.

But I wasn't always an orthodontist. As a teenager, I loved music, math, and physics, and decided to start at UCLA as an engineering major. It was fascinating, but engineering didn't feel quite right for me. I wondered how I could combine my love for science, art, and people, and somehow use the hand dexterity I'd developed from my musical training. Orthodontics was the perfect solution. I transferred to USC and graduated with dual degrees in orthodontics and craniofacial biology.

I find profound joy in creating something beautiful that lasts forever. That's why I love writing and composing music. And as much as I'd like to think of myself as a scientist, I have to admit that it's the art of orthodontics that I'm most passionate about. It's been a privilege to help create smiles and faces that truly reflect inner beauty. I've seen countless patients enter my office as shy, unsmiling children or teenagers, and leave as confident young adults who go on to achieve tremendous personal and professional successes.

Our goal at The Super Dentists is to create extraordinary experiences for our patients. We want to make it fun, comfortable, and easy—not boring, painful, and difficult. We believe if you can look forward to going to the dentist, you'll go to the dentist. We began as a pediatric practice, and we built our whole philosophy and approach around kids and what they love.

In this way, we have removed the barrier of fear. Some of our amenities include theme park–like offices with superheroes, games, dress-up areas, toy stores, and huge slides from one section of the office to another. We also provide complimentary childcare services, tasty beverages from our cafés, custom oral care products, and even augmented reality experiences and original movies starring our very own superheroes and super villains!

Today, our practice treats children and adults in all phases of life, and we have witnessed firsthand the connection between poor oral health and poor physical health. As a result, we've become passionate about addressing this systemic, damaging problem. We want to save as many lives as we can by saving as many mouths as we can. And we want to help kids by helping their parents, because a healthy parent is the number one thing a kid needs.

We have also taken on the coordination-of-care problem by building a group practice with world-class pediatric dentists, orthodontists, general dentists, and other specialists all working alongside each other. We work with the latest technologies and techniques. We collaborate and learn from each other's knowledge and experience. Having all that expertise in one place makes us better at what we do. Maybe someday we'll have physicians on staff as well, creating a new medical model for the future.

Nine years ago, I became a parent, and it was an awakening. The experience of my wife's pregnancy, childbirth, and breastfeeding, followed by the baby's sleep issues, speech development, and all the other aspects of parenting, has been humbling. It has fueled my passion and admiration for children and parents. And it has helped me to think through how to build oral health into a lifestyle from the start. I have built my practice to support that mission.

Through the years, I've heard the same questions over and over.

"But, Doctor, aren't those just baby teeth?"

"But, Doctor, isn't that just a cavity?"

"But, Doctor, isn't this just vanity?"

"But, Doctor, isn't my child too young?"

"But, Doctor, aren't I too old?"

The answers to these and many other questions are found in this book.

Some of the most common questions I've been asked over the years have been related to oral care products. "Is fluoride safe?" "Which toothpaste should we use?" "How do I choose a toothbrush?" "How do we make the kids brush every day?"

This is why we decided to launch SuperMouth™—a revolutionary approach to oral care—because there was nothing in the current marketplace that we could recommend to our patients. Nothing had the safety, efficacy, and fun factor that we wanted for our families. With 25 years of experience in dentistry, we were in an ideal position to create a completely new way to care for mouths. So, we got to work, researching and designing products with safe, effective ingredients that are also fun to use and that promote positive associations with oral care.

From toothpaste to toothbrush, and from floss to mouthwash, SuperMouth is for discerning parents who simply want the best for their children. With playful, superhero-themed products specially designed for every age—from babies and toddlers to young children and teens—SuperMouth offers a comprehensive line of oral care products that work together as a system that not only does the best job of caring for kids' mouths but also is fun to use and builds healthy habits for life.

My education, over two decades of experience, and my daily interactions with other specialists have given me a unique perspective on the power of oral health and its immense impact on the body and the mind. The connections between our mouths and our physical, psychological, social, and emotional health, as well as our longevity, and even our personal and professional successes, are undeniable. There is so much scattered information, and the system is so fragmented, I felt that it was time I put all of this in one place. It was time to connect the dots.

Over the past 25 years, I have watched parents put their children first. This book is for the parents. I hope it helps you take better care of yourselves and your children and have happier, healthier, and much longer lives.

INTRODUCTION

This is not a book about brushing and flossing.

It's about a global health crisis affecting 3.58 *billion* people—nearly half of the entire human population (and no, it's not a viral pandemic).

Let's start with a few facts:

- A 2018 Harvard University study showed that maintaining five healthy habits during your adult life could add 12–14 years to your life expectancy. Can you guess what these five habits are? Eating healthy, exercising regularly, maintaining a healthy body weight, not smoking, and not drinking too much alcohol.

No surprises there. However . . .

- Did you know that having a healthy mouth can increase your life expectancy by up to 10 years?[1]
- And that a healthy airway can increase life expectancy by up to another 15 years?[2] Which means that the health and development of your mouth can actually affect your

longevity more than all of the go-to healthy lifestyle choices we've always known about . . . *combined*!

The fact is that oral health is intimately connected to chronic, systemic disease. This includes many of the "big killers" we all fear: cancer, heart disease, diabetes, obesity, Alzheimer's, and many others. We blame diet, environmental toxins, genetics, and sedentary lifestyles for these conditions. Surely those are all important factors, but have you ever heard someone connect these diseases to poor oral health? Probably not.

But they are intimately and directly connected.

Oral health matters because overall health is affected by it. By "oral health," I mean two things: the balance of microbes that live in the mouth, which can either wreak havoc or support wellness, and the healthy growth and development of the structures of the mouth, which result in good airways and well-formed faces. By "overall health," I mean physical health, social well-being, mental health, success in life, and longevity.

Already, you might be thinking, "But wait a minute! I brush my teeth every day! Twice a day! I floss too . . . well, sometimes . . ."

If that is the case, my compliments—you are off to a good start. But again, this is not a book about brushing and flossing. Nor is it about any of the other general recommendations you can get at your dentist's office. Those are important things we should all be doing, but if you're not doing them, it's not all your fault. You, your family, your friends, and everyone else are victims of a broken system that fails to educate us, fails to provide access to care, and fails to connect medical and dental expertise to recognize and appropriately treat disease. You are not the problem. But you can be a big part of the solution for yourself and for those you love.

This book is another part of the solution. In the following chapters, I will tell you about all of the *profound* ways your mouth is impacting your life and the lives of your children. In addition to affecting your overall risk of developing a devastating chronic illness, your mouth impacts your happiness, prosperity, behavior, mental health, sleep, and longevity.

By learning how and why your mouth has so much power over your life, you can harness that power for good! You can live more than a decade longer. You can find more confidence and happiness. You can sleep better. And you can give your children better lives and brighter futures.

Your mouth impacts everything, often silently, and without you even noticing. Once you start to be more aware, you'll realize that taking care of your mouth is essential in order to have a good life. Your priorities will shift. You'll equate skipping brushing with skipping eating or neglecting your medication. You will also understand how to dramatically lower your risk for many of the most dangerous and debilitating diseases that afflict our species. I am so excited to share this information with you—to arm you with the knowledge you need to be healthier and happier!

THE HEALTH CRISIS IN AMERICA

Many of the oral health challenges we face as adults began in childhood, and these problems continue to plague today's kids. Many children are not getting the care they need for the healthy development of their mouths, bodies, and minds. This is not because parents are neglectful. It's because nobody is educating the public about the mouth-body connection. It's because medicine and dentistry are two distinct fields with very little overlap. And it's because our system of medical insurance often doesn't cover dentistry. Many people simply lack access to the care they need.

Dental caries, the disease that causes cavities, is the most common chronic disease of children. Childhood tooth decay is four times more common than early childhood obesity, five times more common than asthma, and twenty times more common than diabetes.[3] Cavities are so common that they're considered normal.

They're not.

Even in California where I practice—where we pride ourselves on leading the nation in so many health-related areas—many children suffer from undiagnosed, untreated dental issues. By the time these youngsters reach kindergarten, more than half have dental decay. According to the Centers for Disease Control, a whopping one out of five children have rampant decay on seven or more teeth. By third grade, more than 50 percent have had cavities, and nearly one-fifth have untreated decay. Plus, children in California miss 874,000 days of school every single year due to dental problems.

Sadly, adolescents aren't any better off. Fifty-nine percent have had cavities in their permanent teeth, and 20 percent have untreated decay. Toothaches lead to other troubles, like difficulty eating and sleeping, which are likely to impair children's physical development and academic achievement. As you can already see, mouth health quickly becomes whole body health.

Of course, this epidemic extends into adulthood. A full 9 out of 10 people over the age of 20 have some degree of tooth or root decay. Other mouth-related issues include dental crowding, gum disease, and airway-related conditions like sleep apnea—a disorder in which breathing repeatedly stops and starts during sleep.

While the dental crisis is happening on one side, there is an epic avalanche of chronic disease on the other: dementia, obesity, ADHD, anxiety, depression, autoimmune disorders, cardiovascular disease, cancer—the list goes on and on. The average life expectancy in the U.S. dropped three years in a row between 2015 and 2018, making it the worst continuous decline since the period from 1915 to 1918, when World War I raged and the Spanish influenza pandemic killed 675,000 Americans and around 50 million people worldwide.

To complicate matters further, nearly 60 percent of adults in the U.S. have a chronic disease. Four in ten adults have two or more. And chronic illnesses compose seven of the top ten causes of death.[4]

As Americans age and habits don't change, predictions become increasingly dire. For example, more than 6 million people are living with Alzheimer's disease, and by mid-century, this number is projected to rise to nearly 13 million. Between 2000–2019, deaths from Alzheimer's have increased an incredible 145 percent.[5]

If chronic disease is such a big problem, and we know oral disease is a big part of that equation, why has it taken *so long* to connect the dots? The main reason is the historical separation between dentistry and medicine.

This separation began back in the 13th century, when French barbers were responsible for surgical care, including pulling teeth and oral surgeries. Eventually, barber-surgeons would split into two groups: surgeons who were trained to perform complex operations (doctors), and barber-surgeons who performed hygienic services like cutting hair, shaving, bloodletting, and tooth extraction.[6]

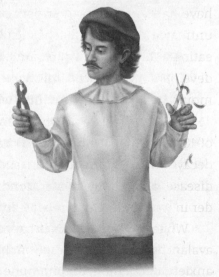

Somehow, removing teeth wasn't considered true medicine and, though it was common for people to die from dental infections, it would be centuries before oral hygiene was considered critical for good health.

Medicine and dentistry almost reconnected in 1837, when Chapin Harris, the man considered to be the father of American dentistry, nearly managed to integrate dentistry into the University of Maryland School of Medicine. But his attempts met with staunch resistance, and three years later, Harris and his partner, Dr. Horace H. Hayden, founded the Baltimore College of Dental Surgery, the first in the world. The independent nature of the Baltimore College of Dental Surgery created a template for other dental schools to follow, and so, dentistry and medicine continued to go their separate ways—evolving as entirely separate fields into the modern era.[7]

CHEW ON THIS

One remnant of the dental/medical separation is the difference between dental and medical attire. To this day, barbers and dentists wear short coats, while physicians wear long coats.

Today, medical doctors ask you to say "Ahhhhh," then look right past your mouth into your throat. When doctors do recognize a problem in the mouth, they often assume your dentist will take care of it. They don't follow up to make sure that you seek the appropriate care, nor do they communicate with your dentist about the physical conditions that might be connected to the oral problems. In other words, there is no coordination of care, even though many diseases of the rest of the body are caused by oral bacteria entering the bloodstream.

On the other side, how many systemic diseases do dentists miss? Signs and symptoms often appear in the mouth first, like thrush, which is indicative of depressed immune function, bad breath from stomach ulcers, or sores from cancer. How many people are suffering from this lack of coordination between the two branches of health care?

In addition to this split, or perhaps because of it, dental insurance is not included in most health insurance packages. And many people can't afford to purchase *more* insurance on top of the prohibitively expensive plans they already have. To make things worse, about 49 million Americans live in "dental professional shortage areas."[8]

Medicaid (Medi-Cal or Denti-Cal in California) covers dental care for some low-income families, but most dentists can't afford to provide it. The payments they receive from Medicaid are far lower than payments from private insurers, so they don't offer appointments to Medicaid patients. In fact, 80 percent of the states in the U.S. blame the shortage of dentists who accept Medicaid for why so few low-income children receive dental treatment.[9] Even people who have coverage can't get seen. As a result, roughly one-third of Americans face significant barriers to obtaining dental care.

Finally, on top of these consequential obstacles, people are terrified of the dentist. Going to the dentist is second only to public speaking when it comes to people's biggest fears. So, even when they have insurance or can afford to go, they still avoid the regular visits that could save and improve their lives.

Oral care *is* medical care. It's a basic need and should be a human right—first, because it can help prevent so many deadly diseases and save lives, and second, because it can dramatically improve the *quality* of the lives it saves.

Think for a moment about the importance of a winning, beautiful smile. It's not just superficial: smiles are cultural indicators. They evoke trust, they inspire attraction, and an unhealthy or unattractive smile is heavily stigmatized. Think about all the associations we take for granted. We associate an underbite with meanness. We call someone a "mouth-breather" when we think they're unintelligent or "unevolved." Your mouth says so much without saying anything at all.

According to the American Dental Association, because of poor oral health, more than one in four U.S. adults avoid smiling and 29 percent of low-income people believe that their appearance affects their ability to interview for jobs.[10]

It makes sense. In America, we equate bad teeth with bad character. We see it almost as a moral failing. We don't associate bad joints or lost limbs with a person's character, but teeth are held to a different standard. Here is another way in which a healthy mouth creates a healthy person and a good life, only in this case it's a psychological and social benefit.

In the following chapters, we will give you a clear picture of *why* oral health is so crucial for your well-being. From there, we'll explore each connection point, starting with how oral health impacts fertility, pregnancy, and childhood. We will examine the connections between oral health and airway and sleep problems. We will look at the psychological impact of an unhealthy mouth. And, lastly, we will examine the physical impact of poor oral health. By the end of this book, you will have a complete overview of how and why your oral health matters, so you can make the best choices for yourself and your family.

You have nothing to lose . . . but your teeth . . . and your health . . . and your life.

Okay, you have everything to lose.

And everything to keep.

Your mouth tells a story. It's the story of your health, prosperity, and longevity. But no matter what your story is today, it's not finished. If your mouth could talk, it would tell you to pay attention, keep reading, and write your own happy ending.

CHAPTER 1

HOW YOUR LIFE AFFECTS YOUR MOUTH

IN THIS CHAPTER

- The critical role of the mouth in whole body health.
- The surprising effects of genetics and epigenetics.
- The role of the microbiome.
- How you can dramatically improve your oral health.

As Americans, we have so many things to be proud of—the electric light bulb, the automobile, space travel, lasers, personal computers—the list goes on and on. But despite our collective genius, we've really managed to fall behind when it comes to health. We rank at or near the bottom in indicators of mortality and life expectancy compared with 35 other developed or developing nations—all while continuing to spend the most of anyone on health care.[1]

We're anxious, depressed, and stressed. We have some of the worst quality sleep in the world, with 1.2 million workdays lost annually to sleep deprivation. We suffer from a myriad of chronic diseases, from Alzheimer's to diabetes. We rank sixth in the world for greatest number of preterm births. And out of 37 member nations of the Organization for Economic Co-operation and Development, the U.S. ranks 33rd in infant mortality.[2] You're getting the picture.

As we have discussed, we are doing no better in the dental realm. The most widespread chronic disease of all, dental disease, now afflicts more than 18 million American children—one-quarter of the population of kids.[3] Teeth are supposed to last a lifetime, but among adults over 65, nearly one in five have lost *all* of their teeth. Dental crowding, gum disease, and airway disorders are on the rise too. And with breathing and related sleep problems come heart, brain, and other body-wide diseases and disorders.

THE GATEWAY TO THE BODY

How is all of this related? It all comes back to your mouth. Nearly everything has to go through your mouth to get to the rest of you, from food and air, to bacteria and viruses, to environmental toxins. A healthy mouth can help your body get what it needs and prevent it from harm—with adequate space for air to travel to your lungs, and healthy teeth and gums that prevent microbes from entering your bloodstream. From the moment you are conceived, oral health impacts every aspect of your life and that of your children.

What happens in the mouth is usually just the tip of the iceberg and a reflection of what is happening in other parts of the body. Poor oral health can be a result of systemic disease, or the cause of it. The microorganisms in an unhealthy mouth can cause not only cavities and gum disease, but also chronic inflammation. And they can enter the bloodstream and travel anywhere in the body, posing serious health risks. Similarly, an underdeveloped mouth can not only result in dental crowding and bite problems, it can also obstruct the airway and interfere with breathing and oxygen intake, potentially affecting every cell in the body. Airway obstruction can also cause sleep-disordered breathing, with health consequences ranging from obesity to cardiovascular disease.

There are many factors responsible for the epidemic proportions of many chronic illnesses in America, but you're about to see the critical role that the mouth plays in your health.

MEET YOUR MOUTH

Hello, mouth! How did you get that way? Let's take a closer look at the gateway to your body.

The Maxilla and Mandible

As you probably know, jaws are the two bony structures that frame the mouth and hold the 20 baby teeth and, later, the 32 adult teeth. The upper jaw, called the maxilla, is formed from the fusion of two bones. The timing of upper jaw fusion, which starts at about age eight, is critical for the timing of orthodontic treatment, as you will learn later.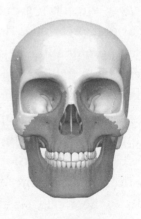

The maxilla has been described as the architectural key to the face because every bone in the face, except the mandible, is in direct contact with it. That's one big job for one (not so) little jawbone! It's important to know that the maxilla doesn't just form the palate and hold the upper teeth. It also contains the maxillary sinuses. The upper surface forms the walls and floor of the nasal cavity and the floor of the orbit for each eye, and it shapes the cheekbones.

Consequently, the maxilla has a profound influence on the shape of the nose,
lips, teeth, cheekbones, and eyes, and on
how balanced (or unbalanced) your face
looks. Since the maxilla is so intimately
connected to so many other facial structures, its proper growth is critical. A poorly
developed maxilla can cause obstructed
airways and interfere with sleep; cause

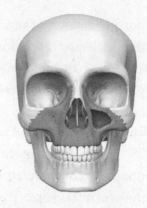

crooked teeth, bite issues, and impacted wisdom teeth; and even
affect eyesight.

The lower jaw—the mandible—also plays an important role in
the proper functioning of the mouth. It is the largest bone in the
face and the only one that is movable. Its ability to pivot in almost
every direction is important for chewing, speaking, and making
facial expressions. The mandible is connected to the skull by the
temporomandibular joints (TMJ). Proper growth of the mandible is
crucial in the development of the dentition, the bite, and the support structures that control the tongue and the airway.

The mandible impacts the lower
third of your face: the chin, fullness
of the lips, proportionality, and balance. An underbite or overbite due
to an incorrectly sized mandible can
have a serious impact on one's overall appearance, and the throat length
(the distance between the chin and
the neck) can add or take away years
from the face's apparent age.

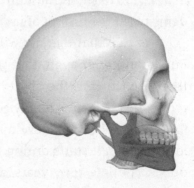

There could be an evolutionary purpose behind why we find
healthy mouths and balanced facial proportions attractive. In this
case, and in many cases, attractiveness and health go hand in

hand. We may not realize it in the moment, but when we seek out one, we're also seeking out the other. It makes good sense that we would find health attractive. We're hardwired to want to have healthy babies, and healthy babies start with healthy parents.

The Growth of the Jaws and Occlusion (Bite)

One of the most common misconceptions I have encountered in my practice over the years is parents' presumption about the growth of the jaws. They assume that, as their child grows, more space will materialize in that child's mouth. This seems like a commonsense assumption: As you grow, things get bigger, right? Not exactly, at least not when it comes to jaws.

When a mom (sorry, dads, it's usually a mom) brings an eight-year-old to the dentist, and the child's front teeth are really crowded, the mom usually assumes we should just wait for the jaws to grow and make room for the teeth. When a mom brings in her four-year-old with an underbite, she assumes we should wait for the kid to grow out of it. You might have guessed by now, but when it comes to orthodontics, waiting is rarely the best approach. If you didn't know that, don't feel bad; many physicians and even dentists get the timing of jaw growth and tooth eruption wrong.

It is true that the jaw will continue to grow as a child grows. But, strangely enough, that growth will not create more space for the teeth. In fact, the space for teeth will get smaller as the jaw grows. If you're now totally confused, let me explain.

The growth and development of dental occlusion is a very long process. It starts during week six of intrauterine life and ends during the late teen years or shortly after 20 years of age. The growth of the jaws and teeth is influenced by both genetics and environment. Before birth, it is primarily influenced by genetics. Postbirth, it is affected by environmental factors such as breast-feeding, mouth-breathing, oral habits, tongue posture, swallowing patterns, diet, early loss of baby teeth, and a number of other dental and orthodontic factors.

I'm going to use the lower jaw to discuss growth and space. That's because the upper jaw has a midline suture (where two bones connect), which doesn't start closing until around age eight, so, when more space is needed, we can easily expand it until that age. However, the lower jaw's midline suture fuses by age one. So, when it comes to space, we are always more concerned about the lower jaw since what we can do there is more limited.

The mandible expands like a growing letter V. A small child's mouth is like a lowercase v, which eventually grows into an uppercase V. As the teeth are transitioning between baby teeth and permanent teeth, the jaw grows backward to make room for first, second, and sometimes third permanent molars. This growth is commonly insufficient to make enough room for the third molars (wisdom teeth) and they get impacted.

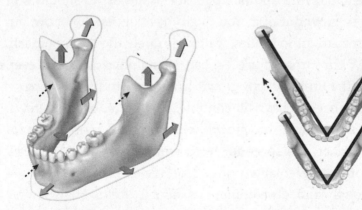

In contrast to the front area of the mouth where permanent teeth are bigger and wider than the baby teeth, the back baby teeth are actually wider than the permanent teeth that will replace them. So, as the back baby teeth fall out and get replaced by smaller permanent teeth, the permanent molars behind them drift forward to occupy this "extra" space, called the "leeway space," decreasing the total arch perimeter.

When it comes to creating space for crowded teeth, orthodontists typ-ically have three tricks up our sleeves: we expand the jaw, remove some teeth, or slen-derize teeth (make them narrower).

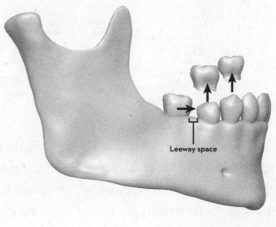

Leeway space

We can do true skeletal expansion only in the maxilla because of the suture I mentioned earlier. Removing permanent teeth can result in a less-than-ideal outcome, both aesthetically and func-tionally. Aside from a few exceptions, avoiding extractions of per-manent teeth can generally help maintain the fullness of the lips, reduce the speed of aging and the development of wrinkles, create a nicer profile, and leave more room for the tongue and airway. Slenderizing teeth can not only create aesthetic concerns (raise your hand if you like tiny little teeth); it may also create a mathe-matical size problem where teeth don't fit together. Plus, excessive removal of enamel from the sides of teeth can potentially jeopar-dize tooth health.

There is a fourth way to make room. It's wonderful, but it requires intervention during a specific window of opportunity before the back baby molars have fallen out. We can place a space maintainer on the permanent molars to maintain the leeway space, prevent those molars from drifting forward, and use the space for the front teeth. This extra space can be significant: around five mil-limeters in the mandible and four millimeters in the maxilla.

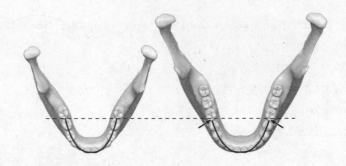

When do you think a child has the most room for teeth? You might think the older they are, the more room they have. Actually, the maximum arch perimeter (the space from the first molar on one side to the first molar on the other side) in the mandible occurs around age eight, and a little later in the maxilla. If left alone, the arch perimeter will actually continue to get smaller and smaller as the person matures.

Maxillary Suture

This suture between the bones of the maxilla begins to close around age eight (with some variation). If a child has crowding, a narrow upper jaw, a narrow airway, crossbite, or issues with snoring or sleeping, the time to address it is before the sutures interlock. An orthodontic expander is typically the best device in this case. It creates more space for the teeth and helps with the growth of the jaw in all three dimensions in space, so the jaw can properly develop. An expander can also help with open bites, underbites, overbites, and overjets. Not only is the result a more beautiful face, but patients can close their lips naturally, breathe easier, and enjoy better health.

Now you can see why the American Association of Orthodontists (AAO) recommends that every child see an orthodontist no later than age seven. By that age, we usually have time to do something about jaw growth issues, and enough space to work with before it

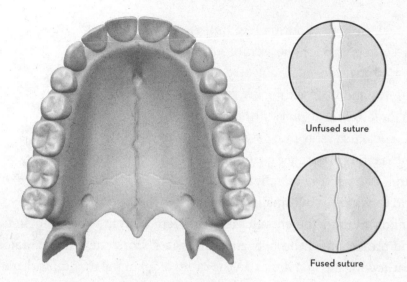

Unfused suture

Fused suture

gets naturally lost. The longer we wait, the greater the chances of wear or other irreversible damage to teeth and surrounding structures. If there are poor oral habits, like thumb-sucking or atypical swallowing, early treatment can be more effective before too much damage has been done. Additionally, there is a psychological benefit to correcting bites early on. The earlier these problems are addressed, the less likely the child is to become self-conscious about his or her mouth, and the less likely he or she is to be teased or singled out by peers.

I should reiterate that age seven is the absolute latest you should see an orthodontist for your child. As you will learn in the following chapters, problems like tongue restriction, oral habits, and airway obstruction must be addressed much earlier. It's also important to note that girls typically go through their growth spurt two years earlier than boys.

Your Tongue

If you are the parent of a young child, you're probably quite used to seeing that child's tongue (you've probably had to tell them to put it back in their mouth a few times). Of course, the tongue is more than something you stick out at people. The tongue is an extremely muscular organ, and it plays a major role in speaking, chewing, swallowing, and tasting. Aiding the tongue are three pairs of major salivary glands and hundreds of smaller salivary 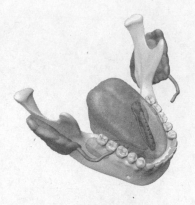 glands around the mouth. They secrete saliva, which moistens and protects the mouth, begins to break down food, and facilitates swallowing.

Your Teeth

As you've already learned, your mouth is much more than just a collection of teeth. It's also important to note that your teeth are more than just inorganic objects that require only brushing from the outside twice a day and cleaning by the dentist twice a year.

Teeth have living internal structures that require nutrients to grow and develop correctly and to help protect them from the harsh environment of the rest of the mouth.

Enamel is the outer layer of your teeth and the part we see. It is mainly composed of the mineral calcium phosphate, arranged in a strong crystal structure

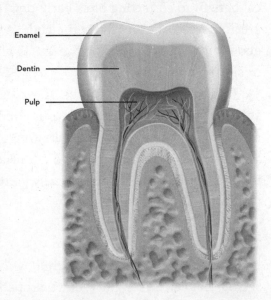

Enamel

Dentin

Pulp

called "hydroxyapatite," which is the hardest substance in your body, even harder than bone. Enamel is the only part of the teeth that does not have living cells.

Dentin is the middle layer and has a team of cells that help to protect the pulp from oral microbes and repair any damage. The pulp supplies the blood and the nervous system that the teeth need for growth and maintenance. Cementum covers the root and is not exposed in mouths with healthy gums. Your teeth are held in place by supporting structures called the periodontium that include cementum, the periodontal ligament, the alveolar bone, and the gums (gingiva).

There are also billions of microbes that call your mouth their home, collectively known as your "oral microbiome."

THE MICROBIOME

Microbes are everywhere on earth. Literally. They are at the bottom of the ocean. They are in volcanic vents, living on sulfur instead of oxygen. They are in the air and water and in every corner of this planet. Microbes are just about everywhere in your body, too, living on your skin, in your gut, and yes, in your mouth.

Your microbiome is made up of all the tiny organisms in and on your body. It's your microbial community, composed of bacteria, archaea, viruses, fungi, and protozoa. And, like a fingerprint, everyone's collection of microbes is different.

Launched in 2007, the Human Microbiome Project has taught us that there are more microbes living in and on us than human cells in our bodies. And our human genes are outnumbered 360 to 1 by our microbial genes.[4]

The thought of carrying around trillions of microscopic critters might make your skin crawl, but before you drop this book to run to take a shower, remember that these microbes are vital to your health and well-being. The human microbiome has evolved alongside us for millions of years and it is key to our survival. Most of your microbes are your allies. They help digest your food. They keep your skin clear by eating dead cells. They work with your immune system to keep you healthy. And they can keep your mouth and teeth healthy. When "bad" microbes want to invade your body, your "good" microbes fend them off. Disease occurs when there is an imbalance in that microbial community, or when microbes gain access to parts of the body where they don't normally live.[5]

Each person houses a unique microbial ecosystem balanced in its own way. This can explain why two genetically identical people—identical twins—can eat the same foods and exercise the same amount, yet they lose or gain different amounts of weight. Or, they can eat the same things and brush their teeth the same amount, and one will get cavities while the other doesn't.

So, where do we get our microbiome? A newborn baby gets its first dose of microbes passing through the birth canal. The vaginal microbiome has evolved to make this just the right healthy starter pack for the baby. Breastfeeding plays an important role as well: the infant not only gets additional microbes from breastfeeding, but breast milk contains prebiotics for certain microorganisms (the food those microbes eat) and antibodies for others (to equip the baby's immune system to fight dangerous organisms). The baby's body gets a crash course on both "good" and "bad" microbes.[6] Then, of course, that baby picks up microbial passengers from other caregivers and the environment, every day.

Different species of microbes live in different parts of the body, forming their own habitats, starting with the mouth. As the gateway to the body, the mouth sends microbes to your gut every time you swallow. The microbiome in the gut has an important job. It helps with digestion, regulates metabolism, and helps your immune system fight off infection. If you have poor oral health, the oral microbiome can get out of balance, compromising that entire system. This can result in serious consequences, including diseases such as colon cancer, diabetes, and rheumatoid arthritis.[7]

Now that you know that bacteria aren't always the enemy, consider the fact that we live in a society bent on getting rid of them. We use antibiotics like crazy and overload on antibacterial soap and antiseptic mouthwash in our homes. And it's no wonder that we do. They have been marketed as the epitome of "clean." Slogans like "Clean Hands Are Healthy Hands," "Protect Those You Care For," and "Kills 99.9% of Germs" terrify consumers into thinking all bacteria are bad. Today, our microbiomes are far less diverse than they once were, and this might be behind the surge in diseases like asthma and allergies.

Our awareness of (and obsession with) microbes dates to the 19th century, when a chemist named Louis Pasteur discovered that microorganisms were causing disease. This work was later extended by a microbiologist named Robert Koch. Their discoveries led to the development of *germ theory*—the origin of much of modern medicine. As you might expect, once people figured out that germs were causing sickness, they started trying to kill those germs. Alexander Fleming discovered penicillin in 1928—a miracle cure for previously deadly infections—and for a while everyone thought we'd won the war on germs. But pretty soon, the bacteria started to adapt and evolve, developing resistance to penicillin and to other newly developed antibiotics. We've been in a neck-and-neck arms race with bacteria ever since. According to the CDC, antibiotic resistance is one of the biggest public health challenges of our time. If we're not careful, we may soon be in a post-antibiotic era.

In addition to the overuse of antibiotics and antibacterial products, a reduction in natural births and increase in C-sections, a reduction in the number of children who are breastfed, and changes in our modern diet all contribute to this troubling trend toward an increasingly depleted and antibiotic-resistant microbiota.

Microbiomes are exciting new frontiers for medicine and nutrition. As we learn more about the many different kinds of microbes and how they work together (or against each other) to keep us healthy or cause disease, we'll start to learn how to harness their power for good.

What's exciting about this new area of science is that there are ways to help our microbial passengers. We can introduce new bacteria or eat with bacteria in mind—focusing on the foods that nourish one strain while starving another. We have so much still to learn, but in the future it's likely that we'll move away from antibiotics (that just kill off every microorganism) and toward microbiome maintenance and support.

FROM ORAL HEALTH TO ORAL DISEASE

Your mouth is filled with billions of microbes. In a healthy mouth, this microbiome population is balanced and even helps in the mineral exchange between your teeth and saliva. Saliva maintains the pH balance and creates the right environment for the microbiome to thrive. Saliva also provides nutrients and antibodies to protect your mouth against infections.[8]

In an unhealthy mouth, the microbiome can get out of balance, switching from helpful to harmful. These microbes can produce waste products that can damage the teeth and gums. This tooth damage, otherwise known as cavities, may initially be minor and limited to the enamel, but if not treated can extend all the way into the pulp of the tooth where the blood vessels and nerves are located. From there, the infection can enter the bloodstream and cause systemic damage.

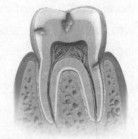

The same can happen with the gums. The bacteria, or their by-products, can cause periodontal disease ranging from gingivitis (which is limited to the inflammation of the gums and is usually reversible), to periodontitis (which involves irreversible destruction of supporting tissues like the bone, cementum, and periodontal ligament).

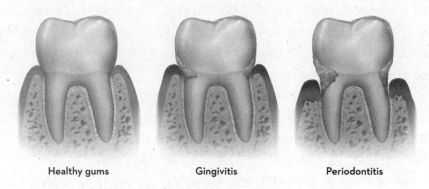

Healthy gums   Gingivitis   Periodontitis

Periodontal disease has been shown to be a significant contributing factor to many systemic conditions, including Alzheimer's, diabetes, heart disease, and others.

All this from bacteria entering the bloodstream through bleeding gums, or from systemic inflammation.

CAVITIES

My patient Laura had never missed a single appointment during the period I was seeing her, but a few years ago, she missed several appointments back-to-back. When she finally came in for a visit and I asked her if everything was okay, she burst into tears and told me she was going through a tough divorce. After such a challenging experience, she had decided to move in with her family in another state to get back on her feet. When she returned to my office a few years later, her mouth was a mess. She had several major cavities and the enamel on her teeth had eroded away so badly that she needed a near full mouth reconstruction. When I discussed this with her, she told me that the stress of her divorce and the pressures from being a single mom had caused her to develop acid reflux. She had treated the discomfort with a constant supply of medications, all while sipping on a steady stream of acidic and sugary drinks like coffee and soda.

Laura was devastated and surprised that she had developed so many oral health issues. After all, she had always been careful about not consuming too many sweets. Additionally, she brushed routinely using a very popular toothpaste specifically marketed to be anti-cavity. What went wrong?

If I ask the average person, "How do you get cavities?" they'll probably say, "By eating a lot of sugar and not brushing my teeth." But if it were that simple, dental caries (or cavities) wouldn't be the number one disease in the world and this section of the book would be much shorter. It turns out, dental caries is a complex process and requires a deeper understanding of all the risk factors involved.

As discussed earlier, everyone has a unique oral microbiome similar to a fingerprint, but it's important to note that, unlike fingerprints, our oral microbiome balance can change over time, for better or worse.

Think of your mouth as a garden and your oral microbiome as the beautiful flowers, plants, and vegetables that grow inside the garden. To keep everything healthy, you water your garden daily, cut and trim branches, and add nutrients to the soil when necessary.

But what happens if there is an issue? What if a weed grows or one of the plants catches a disease? You'd probably selectively take out the weeds or use medicine for that diseased plant and continue taking care of your overall garden. You probably wouldn't throw weed killer on all of your plants. And what happens if you leave town for a few weeks and forget to have someone take care of your garden? When you come back, the weeds may have overgrown and taken over, and diseases may have spread, destroying your crops.

This happens in your mouth too. How well you brush or floss will determine the quantity and quality of your oral microbes, and what you eat or drink and how frequently you do that will favor either the helpful microbes or the harmful ones. Additionally, the oral care products you use can dramatically impact the health of your mouth and ultimately the health of your entire body.

How Cavities Form

Your oral bacteria form a sticky film called "plaque" or "dental biofilm" to attach themselves to the enamel of your teeth. Cavities occur when there is an imbalance of the oral bacterial population and a malfunction of the biofilm. The enamel goes through constant demineralization (losing minerals) and remineralization (gaining minerals) cycles throughout the day. Just about every time you eat or drink, the pH of your saliva drops and becomes acidic thanks to the enzymes that start the digestion of food as well as from the foods themselves. This acidic environment favors the "bad" bacteria. These bacteria digest the food and excrete acid themselves, creating a vicious cycle. The pH of saliva is normally neutral (6.7–7.3) at rest and slightly alkaline when stimulated, but once it becomes acidic and reaches around pH 5.5, some of the hydroxyapatite minerals in the enamel dissolve away from the enamel into the biofilm. When everything is functioning correctly, after about 15–30 minutes, the saliva reverses the pH and bathes the teeth with minerals, which are then absorbed into the teeth.[9]

In a healthy mouth, there is a balance between the demineralization and remineralization of the enamel, and the teeth stay healthy. But if there is a breakdown in that process—for example, if there is too much sugar in the diet or the person is eating and drinking too frequently—the saliva can't put the minerals back fast enough to protect the enamel. If this continues for a long time, the prolonged low pH will cause the oral microbiome balance to shift from healthy to unhealthy and oral disease will take over.

The Infamous Sugar

Ah, sugar. You didn't think you were going to read a book about the mouth written by a dentist without delving into the problem of sugar, did you?

Yes, we all know that refined sugar is bad. In fact, sugar being bad and veggies being good is pretty much all we can agree upon in the medical community. Still, sugar is everywhere. From baby formula, to condiments, to juice, to birthday cakes—the average American consumes more than 150 pounds of sugar every year.[10] That's like eating an entire person made of sugar! Even some well-meaning doctors give lollipops to children after their *medical* visits, although thankfully that trend has been waning.

Why is sugar so hard on teeth? Excessive sugar intake can shift the balance of the oral pH from alkaline to acidic, causing a pre-dominance of bacteria that metabolize simple carbohydrates. These bacteria then release acids that eat away at enamel and cause tooth decay.

RISK FACTORS FOR DEVELOPING CAVITIES

- **Diet**
  Of course, sugar is one of the worst culprits when it comes to enamel care, but it's not the only one. Other refined carbohydrates, processed foods, and acidic foods can all lead to cavities.

It's not just the bad things in our diet that can cause cavities. Oftentimes, it's what's missing from our diets that is the problem. For example, calcium is a key nutrient for healthy bones and teeth. However, without the adequate amounts of vitamins A, D, and K2, calcium will not be deposited where it needs to go, no matter how much of it you consume. Although most Americans get adequate amounts of vitamin A from food, almost everyone is significantly deficient in vitamins D and K2 because of our lifestyles and changes in our modern diets. Vitamin D helps transport calcium from your intestines, and vitamin K2 helps activate the proteins needed to pull the calcium from blood and direct it into your teeth and bones. Without sufficient amounts of these nutrients, your teeth will not be able to grow healthy and adequately protect themselves against damage.[11] (See Chapter 5 for more on vitamins.)

- **pH balance**

  pH is a measurement that allows us to know if a solution is acidic or alkaline. It ranges from 1 (acidic) to 14 (alkaline) with 7 being neutral. What you put in your mouth, and how often, affects the delicate pH balance of your oral microbiome. Eating acidic foods and sugar are just two ways that we disrupt the pH balance, but smoking, drugs, and using harsh toothpastes and mouthwashes also tilt the balance toward the acidic, allowing the bad microbes to multiply. The more alkaline your saliva, the more it favors the "good" bacteria, and the more acidic the pH is, the more the "bad" microbes thrive.

- **Lack of saliva**

  Dry mouth affects about a third of the population, and most people suffer from it at some point. Some causes are mouth-breathing, medical issues, medications, poor diet, and stress.

- **Genetics**

  Even though people commonly blame their ancestors for their oral health problems, only 10 percent have genetic causes. Ninety percent of all cavities are related to the environment or are associated with oral care, which means they are controlled by us.[12]

- **Antibiotics**

  Antibiotics, antibacterial products, and potent essential oils can decimate the oral microbiome.

- **Poor oral care routine**

  Poor brushing and flossing can result in biofilm that becomes too thick, turning into tartar or calculus, which tilts the balance from oxygen-dependent aerobic bacteria to opportunistic and aggressive anaerobic bacteria.

- **Harmful oral care products**

  Many oral care products indiscriminately kill the oral microbiome, disrupting the delicate balance. They may transform beneficial microbes into a pathogenic state or allow new, more opportunistic ones to take hold. The need for this delicate microbiome balance is why I caution everyone against using random, over-the-counter toothpastes and mouthwashes, especially those that contain antibacterial ingredients or alcohol. Please refer to Chapter 5 for a comprehensive oral care checklist and a full list of Mouthstanding™ oral care ingredients and products for you and your children—and a list of Mouthrageous™ ones you should avoid at all costs.

The good news: Just as the oral microbiome can shift from healthy to unhealthy, that pattern can also be reversed with simple changes such as having a great oral care routine at home and maintaining regular visits to your dentist.

The great news: We now understand more than ever about the various mouth microbes and what they do. In the future, I predict we will have individualized probiotic and prebiotic products, designed with your individual microbiome in mind—to support the optimal health of your unique oral habitat.

THE SIMPLE THEORY VS.
THE COMPLEX REALITY

One of the first things I learned when I started to drive was that objects in the mirror are closer than they appear. And one of the first things I learned as a scientist was that processes in the body are a lot more complex than they appear. They are so complex that we try to simplify them so that we can understand. We create a simple theory to explain a complex reality. And in the process of simplifying, we often overlook important information.

Our cells and body parts do not work independently or in a vacuum. There is a tremendous amount of detailed cooperation and collaboration between the functions of our cells and organs. For example, imagine the intricate interaction it takes to play the piano. Your brain sends signals to all ten fingers with ten different instructions to play ten different notes, and then your fingers have to respond at just the right tenth of a second. The workings of cells and organs happen in a precise sort of concert too. They move and grow together. What affects one part affects the whole. Neighboring tissues can produce chemical signals that affect how quickly each one grows, or they push or pull, making each other grow in different directions. The palate is a great example of how neighboring tissues and structures influence each other's growth.

Your palate is the roof of your mouth, and it's also the floor of your nose. Think of it as a two-unit duplex with your mouth on the bottom and your nose on top. The development of your mouth has a direct impact on the development of your nasal cavity. This is why a narrow jaw with dental crowding can also lead to a constricted airway and result in sleep-related breathing problems such as snoring or sleep apnea.

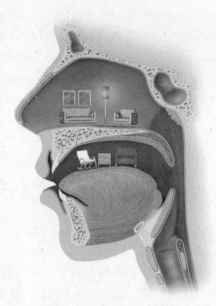

Your tongue also has an important role to play. If it is positioned properly against the palate, it helps the palate grow into the proper shape. As you will learn later, many other factors can impact the growth of the palate, including breastfeeding—or a lack thereof—and whether you breathe through your nose or mouth. Breastfeeding is just one example, but it's a good illustration of how a seemingly unrelated factor can have lifelong consequences, like crooked teeth, impacted wisdom teeth, airway problems, sleep disorders, and a host of other systemic health issues.

When it comes to both growth and function, it's all about complexity and cooperation. Think about how much coordination it takes for your body to carry out the functions you perform every day. For example, chewing requires complex coordination between your tongue, teeth, salivary glands, jaws, temporomandibular joint, lips, cheeks, palate, and brain. Swallowing is incredibly complicated, too, requiring a series of voluntary and involuntary actions and reflexes. But don't think about it too much or you might choke!

This glorious complexity gets overlooked when we approach our bodies with a simplified understanding. It doesn't serve us very well, especially when we expect a "quick fix." For example, let's say that you're having difficulty sleeping. You see a doctor and the simple theory emerges—you have "insomnia." You take a sleeping pill and hope for the best. However, insomnia might be a symptom of a much more complicated underlying issue. Maybe it's the result of gastrointestinal problems, like acid reflux. Maybe you're suffering from sleep-disordered breathing, like sleep apnea, and those sleeping pills are actually dangerous because they prevent your body from rousing itself when it needs oxygen. And while you're missing the big picture, you may be creating more problems, like the side effects of sleeping pills.

When there are problems, like insomnia, they often run much deeper than the obvious symptoms they create. Symptoms are red flags, not problems in themselves.

A NEW TWIST ON THE "NATURE VS. NURTURE" DEBATE

When a new patient comes into my practice, I often hear them say, "My parents had bad teeth, so I have bad teeth too." But our hereditary characteristics have a much smaller impact on our oral health than we give them credit for. In fact, our habits and behaviors make a much bigger mark than any genetic factors. How could that be?

The answer is epigenetics.

But I'm getting ahead of myself.

Let's back up for a minute. The debate over nature (the genes we are born with) versus nurture (lifestyle and environmental influences) has been going on for a very long time. And the more we've learned about how genes work and are expressed, the more complicated the nature vs. nurture picture has become.

Jean-Baptiste Lamarck, a 19th-century French biologist, is famous for the idea that acquired characteristics could be inherited. In other words, Lamarck believed that a person could pass on the traits they gained through experience. For example, he thought you could build up huge muscles by working out and then pass those muscles on to your offspring.

Charles Darwin seemingly debunked that idea when he published his famous *On the Origin of Species* in 1859. Darwin believed evolution was about changes over time by random mutation and natural selection—favorable genes that contribute to an animal's survival get passed on. So, successful animals—ones that manage to reproduce—are the ones who pass their DNA on to the next generation: survival of the fittest. For Darwin, it had nothing to do with acquired traits—big muscles or bilingualism or expert marksmanship—and was all about which genes gave a creature an advantage that allowed that creature to have babies.

Almost a century later, in 1953, James Watson, Francis Crick, and Rosalind Franklin discovered the double helix structure of DNA. It was the first time that people started to understand that this molecule with its mysterious code was responsible for determining how we look and, in many ways, who we become. Following this discovery, science entered a period of frenzy as everyone worked to understand what DNA was doing. In 1990, the Human Genome Project set out to decode the entire human genome.

The Human Genome Project was an immense step forward in our understanding, but it was also something of a letdown. There weren't nearly as many genes as scientists expected to find, and they still could not figure out how DNA can code for so many dizzying variations.

We began with a simple theory. Scientists believed we would find a gene for everything from eye color to depression. But as the genome map emerged, complexity emerged with it. Once we began to understand the blueprint, we realized that the blueprint could change.

GENETICS (HOW DOES THIS STUFF WORK?)

If you aren't familiar with how DNA works, it may feel like an intimidating topic. But, while the intricacies are complicated, the general structure and function of the DNA molecule is quite easy to get your head around.

Say you want to build a house. You'll need a blueprint to give you detailed instructions on where all the rooms are, how big they are, how the plumbing or electrical system is laid out, and many other details. When it comes to our bodies (and the bodies of every other living organism), these instructions are encoded in the DNA molecules inside every cell.

Nearly every cell in your body has the same DNA. Some parts of the DNA strand instruct your body to build a liver, other parts code for the brain, some parts give you blue eyes, and others shape your mouth.

The entire YOU blueprint is present in each of your cells. That's why, theoretically, you only need one cell to clone an entire organism. One cell gives you all the information.

DNA unzips and is copied to form proteins; proteins do most of the work in cells; cells make up tissues; tissues make up organs; and organs work together to form, well . . . you.

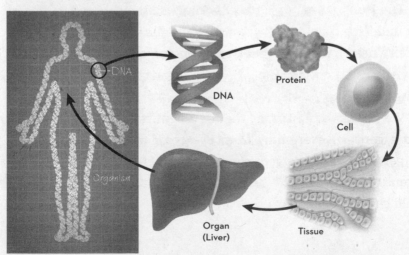

CHEW ON THIS

The structure of DNA—the double helix—resembles a twisted ladder.

The rungs of the ladder are made of nucleotides. Each nucleotide has a backbone and a base. There are only four types of these bases, and each type has a letter: A (adenine), T (thymine), C (cytosine), and G (guanine). You can think of these four bases (A, T, C, G) as four letters in an alphabet or four notes in a musical instrument. DNA uses only four bases over and over again in different sequences to express itself in infinite ways, creating all the species in the world.

Bases play favorites. A always binds with T, and C always binds with G. Each rung of the DNA ladder is a pair of bases that meet in the middle.

The human genome contains about three billion of these base pairs. A typical gene—a unit of heredity—is approximately 1,000 base pairs.

The things DNA "makes" are proteins, which are large but very versatile molecules. Proteins do most of the work in cells and are involved in transport, structure, function, and regulation of the body's tissues and organs.

In order for DNA to instruct the cell to create a protein, the cell has to read the DNA code. So, the twisted DNA ladder "unzips" and the exposed nucleotides bind to new pairs, creating a copy. That copied segment is the code for one particular protein that is built to do something specific and necessary for the body.

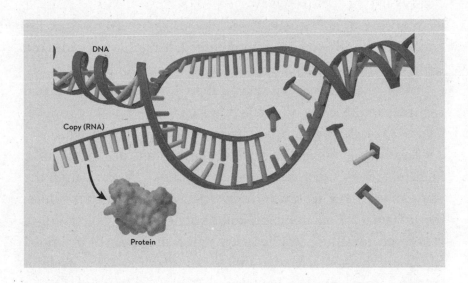

DNA

Copy (RNA)

Protein

EPIGENETICS—WHAT GETS TURNED ON, AND WHERE

You might be wondering, if DNA is in every cell, and it codes for *everything*, how does a liver become a liver and not, say, a tooth or a heart? How does the body determine what information is used, when, and where?

Let's go back to our house analogy. When you need to do the plumbing for the kitchen, you turn right to that section in the blueprint. You probably don't read all the pages of the blueprint from front to back. It's the same with DNA. In the liver, only the parts of the molecule relevant to the liver turn on, while the rest of the molecule sits dormant.

How does the body know *what* to turn on *where*?

There are a lot of factors that affect what turns DNA off or on, in a given location. In some cases, proteins bind to the DNA molecule in specific places, based on a sequence of nucleotides.

There are other things that affect where DNA is copied, too, that aren't just the code itself.

Often, there are compounds stuck to the outside of the DNA molecule. They are little molecular hangers-on, and they can affect what is copied, and where. These molecules can come from the environment, and they can be inherited. They have a profound impact on the expression of genes, for good and for bad.

When these molecules attach, they don't change the actual sequence of the DNA—the code of nucleotides—but they can still be passed on through the generations. They can stay stuck!

This is called epigenetics ("epi" means "on" or "on top of"). Remember how I just said these compounds can come from the environment? The fascinating new science of epigenetics studies the influence of environment and lifestyle on the expression of genes, and how those acquired epigenetic changes can be inherited.

To put it another way: if things can attach to the outside of DNA and change the way it's expressed, then our heredity is about much more than a code we're born with. It's about our environment too! The function of the DNA molecule itself is both nature *and* nurture.

Continuing with our house analogy, a good way to understand how epigenetics works is to think of the DNA sequence as the text in that blueprint to build a house. Epigenetics would be as if someone used a marker to cover parts of the letters or words so they couldn't be read, or highlighted certain parts to emphasize their importance. Again, epigenetic influencers don't alter the genes themselves, but they do change the way those genes are expressed.

Lifestyle choices, oral care habits, stress, exposure to environmental toxins, that cigarette you just smoked, that extra time at the gym—these can affect what sticks to your DNA. And what sticks to your DNA can change how your body functions, and how the bodies of your babies function and look, from hair color to intelligence to the size of their mouths.

Lamarck has been ridiculed in biology circles for nearly two centuries, but in a way, he was correct! Acquired traits *can* be passed on. And it's because of epigenetics.

Here's an example of epigenetics in action: A 2018 study in mice showed that the benefits of exercise may in fact be inherited through epigenetics, at least when it comes to brain benefits.[13] Researchers in Germany found that a physically active father may have smarter children, even if he didn't start exercising until adulthood.[14]

In another study, researchers were able to show that fear can potentially be passed on to future generations.[15] They conditioned mice to be fearful of a particular scent (acetophenone or cherry blossom) by exposing them to the smell alongside a mild electric shock. Eventually, the mice would shake from fear when exposed to the scent, even in the absence of a shock. The offspring of those mice (for two generations!) would also shake when exposed to the scent of cherry blossoms, even though they had never been exposed to the electric shock accompaniment. The researchers even identified the actual brain changes responsible for the scent response.

Here are additional examples that researchers increasingly believe are due to epigenetics:

- Children born to pregnant women who smoke have a greater chance of developing asthma.[16]
- Boys who overeat during their prepubescent years may end up having children and grandchildren with significantly shorter lives.[17]

The meaning of this is truly staggering. Your father's gym habits could have affected your intelligence. Your mother's smoking habit could be affecting your lungs. Your grandfather's lifestyle choices could be impacting your life expectancy. Your parents' fear of dentistry could be making you fearful too.

It's fascinating (and a little disturbing) to think about the potential implications of this revelatory new insight into our genetics. If the environment is affecting how genes are expressed, what *in* the environment is having the greatest impact? There are so many environmental factors to consider, like exercise, smoking, and stress, but another big one, maybe the biggest of all, is diet.

We put food in our bodies every day. Our digestive systems break that food down into the building blocks of our bodies. Our cells are constantly interacting with those building blocks—vitamins, minerals, proteins, carbohydrates, and fats, but also preservatives and all sorts of other laboratory-created compounds.

Here's an example of how nutrition interacts with DNA. One very common epigenetic molecule (one of the most common hangers-on) is a methyl group: CH3. It's a molecule made up of a carbon atom and three hydrogen atoms and it sticks on the DNA molecule like angry on a wet cat. Nutrients such as folate and vitamins B12, B6, and B2 are essential raw materials for methylating DNA.[18]

Methyl groups are essential for the proper functioning of DNA in many contexts. But over- or under-methylation—where DNA has way too many or two few attached methyl groups—can cause all sorts of problems. For example, a lack of key vitamins in one's diet can result in under-methylated DNA. Both under- and over-methylation have been connected with chronic physical disease, mental illness, and premature aging.[19]

WHAT DOES IT ALL MEAN?

Put simply, our diet and lifestyle choices matter a lot more than we may have imagined.

Once again, let's continue with our house analogy. If you use subpar materials, or cut corners during construction, you're going to have problems. The roof might leak, the windows will be drafty, the drywall will crack, and the pipes will burst. Your house will deteriorate quickly, and if you planned to live there, you'll be in trouble.

Your body is just like that house. If you eat subpar foods without enough vitamins, minerals, and other essential nutrients, you're going to have problems. Your mouth might not develop correctly, your body will be weaker, and you will be more likely to succumb to illness and the ravages of age. If you planned to live a long, happy life in that body, you'll be in trouble.

The fact is, we have a long way to go to understand exactly how most epigenetic changes affect our health and our children's health, but we are learning more every day. The most important takeaway is that our habits and lifestyle choices really matter when it comes to the health of our babies. Eating healthy, exercising, and yes, taking care of your mouth, can positively influence your epigenome. If motivating yourself into a healthy lifestyle is hard for you, remember you are not just changing *your* reality—your behavior today can affect the evolutionary cards dealt to your children, grandchildren, and beyond.

THE GIVE-AND-TAKE OF EVOLUTION

In Darwin's theory of evolution, "successful" traits get passed on.

Consider this scenario: A certain area is populated by many hard-shelled beetles, but few other insects. There are birds in the area, but most don't have beaks sharp enough to break the beetles' shells, so they can't digest them. Many of those birds die of starvation before they reach reproductive age. The few birds who do have sharp-enough beaks eat well and go on to create healthy eggs and healthy babies. Those babies grow up, and most of them also have sharp beaks, because their parents did. Sharp beaks get reproductively selected for, and, over generations, all the birds in that area have sharper beaks than their ancestors.

This is how Darwinian evolution works. The environment puts pressure on the animals that live there, and that pressure influences the evolution of those species of animals. The creatures that can survive the challenges have babies. The creatures that can't, don't.

There is often a cost versus benefit when it comes to evolution. When something changes to accommodate something environmental (such as a new food source) or something intellectual (the ability to talk), it often results in losing something else. Changes present their own challenges!

For example, the first *hominins* in Africa, a variety of species that later evolved into humans, started walking on two legs around four million years ago. This adaptation allowed them to thrive in a difficult environment, but the upright pos-  ture combined with downward-facing nostrils meant that air had to take a 90-degree turn as it entered the skull.[20] This disruption in air flow was an evolutionary trade-off—worth it for the added mobility of a two-legged stance, but a compromise, nonetheless.

Another evolutionary trade-off: around two or three million years ago, human brain size exploded to three times the size of a chimpanzee brain (our closest living relative). This led to an increased learning capacity and enhanced motor skills. But, as our brains grew, our faces became more vertical, and our mouths became smaller.[21]

Then, about a million years ago, there was another big shift. The larynx in the throat descended relative to the soft palate.[22] This

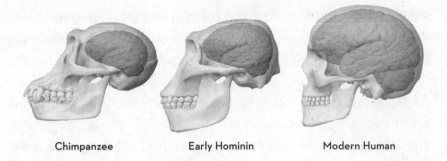

Chimpanzee   Early Hominin   Modern Human

vastly improved speech, something that has allowed us to advance far beyond all other animals. But it also resulted in a rearrangement of parts of the throat and airways, making breathing and swallowing less safe and more challenging than it is for other animals. That's why it's relatively easy for humans to choke, why a drink can shoot out of your nose when you laugh, and why many of us have airway problems like sleep apnea. But, in the scheme of things, those are reasonable prices to pay for the ability to talk.

While our faces may be a bit nicer-looking, the reduction in size of the human mouth has made it more likely that we'll have oral-health problems. And without optimal nutrition and lifestyle habits, these problems have been compounding.

Our mouths have continued to get smaller and smaller, and it's not all because of genetics. The advent of stone tools to cut food and cooking to soften it reduced our ancestors' need for larger jaws. The transition from foraging to farming followed by the Industrial Revolution further expedited the miniaturization of our mouths.

Our modern, soft, sugary, processed foods not only lack the necessary nutrients for the healthy growth and development of the mouth—they are also too easy to eat. When it comes to muscles, you use them or lose them. And our bones respond to mechanical loading during growth. When we eat soft foods, it weakens our jaw muscles and bones, and results in smaller mouths.[23] Before the Industrial Revolution around 200 years ago, impacted wisdom teeth were almost unknown. Today, over 70 percent of us suffer from them. Once upon a time, our larger jaws provided plenty of room for our teeth. Today's children have such crowded teeth that more than four million Americans now wear braces at any given time.

What is alarming is that while genetic changes in evolution can take hundreds of thousands (or even millions) of years, epigenetic changes can happen in a single generation. I see this in my own practice. I'm frequently asked by parents why they have straighter teeth than their children.

HOW IT ALL BEGAN: A BRIEF HISTORY OF FOOD

Here we are, 21st-century America. Our "Standard American Diet," for which the acronym SAD says it all, is a disaster and a significant contributing factor to chronic disease. Processed foods, simple sugars, and empty calories don't support our cells and microbiomes the way they should, and this impedes the healthy growth and development of our mouths and facial structures. Our gut microbes that need complex carbohydrates, fructose, healthy fats, and other vegetable-based nutrients are starving. The sugar we eat upsets the delicate balance in our mouths, too, creating the perfect conditions for destructive bacteria to thrive.

If you're like most Americans, chances are your diet is a combination of cooked, soft, and highly processed food, much of which is significantly modified from the way nature originally intended. While other species spend a significant part of their waking existence acquiring, chewing, and digesting natural food, we take a few steps to the refrigerator, grab what we want, and indulge. As a comparison, a chimpanzee typically spends about half of the day feeding and chewing his food.[24]

The invention of simple food preparation tools approximately 2.5 million years ago was a turning point in the evolution of human diets. These tools were used to cut meat and plants and remove marrow from the bones, and they were also used for pounding and tenderizing tough foods.

At some point, 250,000–700,000 years ago, our ancestors added cooking to food preparation, which was another significant turning point. Cooked food not only tasted better—humans were evolving to appreciate a wide range of flavors due to the way the nose, mouth, and throat were developing—but it also presented some other key evolutionary advantages. Cooking allowed for better digestion of certain proteins and carbohydrates, it removed harmful microbes and parasites from the food, it extended food storage time, and it made food easier to chew. Our prehistoric ancestors probably had to chew at least twice as much as a typical modern human.[25]

Aside from environmental and social factors, these changes in what people ate and how they prepared their food may have provided the additional calories and nutrients that allowed the brain to grow larger. But they were also related to the evolution of smaller faces and mouths.[26]

Up until 15,000 years ago, our ancestors lived as hunter-gatherers, and most of their food was obtained by foraging. Around 10,000–14,000 years ago, many human populations around the world, independently of each other, started experimenting with agriculture. There is a lot of evidence that foragers had stronger bones and teeth than the farmers who came after them.[27, 28]

This agricultural revolution had a significant impact on our civilization, for good and bad. With control of the crop against natural droughts and floods humans could build a food surplus for the first time in history. This meant that not everyone had to be involved in hunting or gathering food, which led to the specialization of labor (cooks, mechanics, writers, engineers, teachers, dentists, etc.) and allowed large population centers to flourish.

But the agricultural revolution also had some unintended consequences. Since we were picking out certain seeds over others—planting the seeds from plants that tasted better or looked prettier—we were inadvertently tinkering with the crops' DNA. For example, about 9,000 years ago, farmers started selectively breeding a wild grass on the American continent known as "teosinte" for size and taste.[29] Over the following few thousand years, that grass transformed into the crop we know as corn.

In another example, selective breeding over thousands of years has changed a tiny, bitter fruit into the giant, red, sweet watermelon on our modern tables.[30] Today, the rate of selective breeding has been accelerated by more advanced techniques like genetic engineering to develop varieties of fruits and vegetables that can ward off pests or withstand drought.

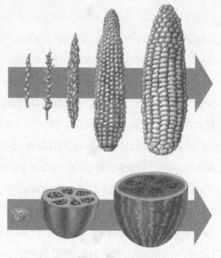

We have also been selectively breeding livestock like cattle and sheep, changing their shape and size to please our palates. In the case of these domesticated animals, we have accelerated the rate of change with artificial growth hormones, antibiotics, and other growth- and development-related treatments.

As we were changing the taste and appearance of plants and animals on the outside, we were also changing their composition on the inside: removing certain nutrients and adding others. For example, in the case of corn and watermelon, we have more than tripled the percentage of sugar in these crops as compared to their natural variations.[31]

With the advent of farming around 12,000 years ago, we significantly changed our food supply in the blink of an evolutionary eye. Meanwhile, humans have not had any significant macroevolutionary changes in 60,000–200,000 years. Our mouths and bodies have not kept up with the rapid changes in our diets, and this has affected our oral and physical health.[32]

Over the past 1,000 years, change has continued to speed up. From the introduction of sugarcane to the Industrial Revolution, food and culture have worked together to change our diets and habits for the worse. The Industrial Revolution forced entire families to move from rural to urban areas, and higher costs and low wages required more women to enter the workforce.[33] As women went to work, breastfeeding became more difficult, and sugar-filled formula took its place. The migration to the cities also required food that could be easily stored and transported without spoilage for long distances. The mass production and distribution of processed foods brought new, easy, nutrient-deficient sustenance to more and more people. During the last 250 years, farming practices and the use of artificial fertilizers, pesticides, and herbicides have depleted the natural fertility of soil.[34] We are inundated with all manner of unhealthy, damaging food and drink: sodas that are almost more sugar than liquid, fast food, microwavable entrees, and everything else we see in the frozen food section of the local supermarket.

It's fascinating to look at how culture has shifted to accommodate the shifts in diet and the subsequent health-related consequences. For example, once upon a time, sugar was a delicacy—something enjoyed only by the wealthy and prominent. As such, tooth decay was a symbol of wealth and prosperity. Queen Elizabeth I was famous for her love of sugary foods and for her decayed, black teeth. During that period, cavities were a rich man's disease (or rich woman's disease, in the case of the queen). People would actually blacken their teeth to appear more well-to-do.[35]

THE EVOLUTION OF CAVITIES

While we were busy tinkering with seeds, altering our food, distributing cola across the continental U.S., and formula-feeding our babies, the bacteria in and on our bodies were busy too. They were busy *evolving*. Human evolution takes a very long time. But bacteria evolve quickly. Often, bacteria produce multiple generations in a single day (a generation takes from about 12 minutes to 24 hours, depending on the organism and the conditions).

One bacterium in particular has evolved in direct response to the rise of human civilization—capitalizing upon our long-standing love affair with sugar. It's called *Streptococcus mutans*, and it's one of the primary bacteria responsible for cavities.[36]

Around 10,000 years ago, the population of *S. mutans* increased significantly. It was about the same time that humans started farming, significantly increasing the amount of carbohydrates in their diets. Over the years, *S. mutans* got better and better at breaking down sugar, and at surviving in the acidic environment of a starch-filled mouth. As you might expect, as *S. mutans* evolved to thrive in the mouths of carbohydrate-fed humans, cavities became more and more common.

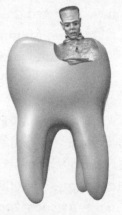

We created a monster!

Today, thanks to our Frankensteinian *S. mutans*, you have to brush, floss, and visit the dentist regularly, even if you eat well.

RESTORING THE HABITAT

Eating as many natural, whole, organic foods as possible is a great first step. Avoiding excess sugar and processed foods will go a long way toward supporting your biology. But we've not only changed the nutritional composition of our foods for the worse—the bacteria we're contending with today are some tough stock too. Today, it takes more than good eating to keep your mouth and body healthy.

It seems the more we learn, the more we realize that there has been much greater wisdom in nature than we ever gave it credit for. Maybe it's time to gain some humility: stop interfering with nature and work *with it* instead. Millions of years of evolution will inevitably create more elegant, balanced solutions than a few centuries of scientific inquiry ever could.

With that in mind, here are a few things you can do to help your microscopic cohabitants:

- Eat as much natural, whole, organic, fermented, and high-fiber foods as possible.
- Decrease consumption of sugar and processed foods.
- Avoid frequent snacking throughout the day.
- Reduce stress, as it may impact bacterial diversity.[37]
- Exercise regularly to boost your "good" microbes.[38]
- Quit smoking.[39]
- Get at least seven hours of quality sleep every night (more on this in Chapter 3).
- Get dirty: play outside, work in the garden, and so on.
- Get a pet to increase your overall microbial diversity.[40]
- Avoid antibiotics as much as possible.
- Avoid antiseptic cleaning products.
- Avoid antibacterial soaps.

- Don't use harsh toothpastes and mouthwashes (more on this in Chapter 5).
- Take care of your mouth.
- Parents-to-be: consider natural (vaginal) birth and breastfeeding.

CONNECTING THE DOTS

Eating well, sleeping well, breastfeeding, avoiding antibiotics whenever possible, visiting your dentist regularly, and practicing good oral hygiene are all ways you can protect your health, the health of your children, and the health of future generations. So much of your story really is in your hands.

CHAPTER 2

PREGNANCY, INFANCY, AND EARLY CHILDHOOD

IN THIS CHAPTER

- The crucial role of a healthy mouth in conception and pregnancy.
- The reason why baby teeth are necessary (and how to care for them).
- The case for breastfeeding.
- Tips on how to keep your child's mouth in tip-top shape.

Getting ready to have a baby can be a time of excitement and joy: trying to get pregnant, then waiting for your baby to arrive, guessing the sex of the baby, picking out names, decorating the baby's room, and daydreaming about what they will be like. For many people, it's one of the most precious times of life, and women tend to work hard to take good care of themselves during this time. They eat well, avoid potentially dangerous foods or medicines, exercise, go to regular doctor or midwife appointments, and otherwise do all they can to ensure the health of their growing child.

However, while expectant mothers are told not to drink or smoke, to avoid lunch meats and mercury-containing fish and soft cheeses, they aren't made aware of the importance of caring for their mouths. It's one of many examples of a situation in which oral health is overlooked or underplayed. It might be the most important example of them all because it directly affects the health and safety of both mother and baby.

We've known for over a century that you can't have a healthy body without a healthy mouth. Now, we are learning that you may not be able to have a healthy pregnancy without a healthy mouth, either.[1] Sometimes, an unhealthy mouth means you can't even get pregnant in the first place.

BEFORE PREGNANCY

If you and your partner are thinking about getting pregnant, you should both go to the dentist. This is good advice for anyone, but especially if you know your oral health isn't the best. Why? Because unhealthy gums may cause infertility! A recent study showed that the presence of a common periodontal bacteria in saliva that indicates poor oral health was three times more prevalent in women who did not become pregnant and was associated with a significantly increased risk of infertility.[2]

This isn't just important for women; it's important for men too. Gum disease can lower sperm count and negatively impact sperm quality and motility.[3] And the bacteria in the mouths of both parents can be transferred to their baby after birth. (More on this later.)

DURING PREGNANCY

There may be a good reason why periodontal disease prevents pregnancy: it's dangerous for the developing fetus. Periodontal infection has been linked to several medical conditions relating to pregnancy, including premature birth and low birth weight, both of which can have short- and long-term health consequences for the baby.[4] Periodontal infection—or gum disease—is a serious gum infection that can not only result in dental problems but can introduce bacteria to the bloodstream with whole body consequences.

Maternal oral health affects both the mother and her developing baby. Numerous studies report that poor oral health raises the mother's risk for gestational diabetes and preeclampsia, a dangerous pregnancy complication.[5, 6] And oral health extends to airway health. In fact, up to 46 percent of all pregnant women snore. And women who snore habitually are three times more likely to deliver developmentally delayed babies.[7] Sleep apnea tends to worsen during pregnancy as well, from weight gain and the baby pushing against the diaphragm. Obstructive sleep apnea (OSA) is also associated with health risks for both mom and baby and is another problem an airway-focused dentist can help address.

Pregnant women with obstructive sleep apnea have a four-times-higher risk of hypertension, and twice the risk of gestational diabetes. Babies of obese pregnant women with sleep apnea are about twice as likely to need a C-section and three times more likely to need neonatal intensive care, and about 75 percent of babies born to moms with OSA have lower birth weights and lower Apgar scores (a test used to assess a baby's health soon after birth).[8] All of this makes sense if you think about what obstructive sleep apnea does—it limits the amount of oxygen in your body. You literally can't breathe! Of course, this would have detrimental effects on a mom and her developing fetus.

Pregnancy is a tough time to care for one's mouth. Nausea and vomiting may make brushing difficult. The acid from vomit may damage teeth directly, especially the top front teeth. It may seem counterintuitive, but pregnant women should avoid brushing right after vomiting, since that's when the teeth are the most vulnerable to degradation (I know, it's gross, but a rinse with water is still okay). Pregnancy can bring other unpleasant mouth problems, too, like bad breath (halitosis) and more mobile teeth, especially in the last trimester.[9]

Food cravings, another common pregnancy issue, often mean dietary choices aren't the healthiest—an increase in sugar or carbohydrate consumption can provide tasty food for harmful oral bacteria. And hormonal and immune system changes during pregnancy make pregnant women more susceptible to a common form of gum inflammation called pregnancy gingivitis. Increased bleeding and gum sensitivity are thought to be a result of higher progesterone levels and the effects of pregnancy on the cardiovascular system.[10]

Pregnancy is one of the most important times to see a dentist, and paradoxically, when women tend to visit the dentist the least. In California, only one-third of pregnant women see the dentist during pregnancy.[11] That's half as many as who go during a typical nonpregnant year. It makes sense: between doctor visits for baby checkups, preoccupation over baby preparation, physical discomfort, work demands as women prepare for going on leave—it's an overwhelming time to do just about anything! Who feels like going to the dentist?

To complicate things further, many dentists are reluctant to treat pregnant women because of unclear guidelines or a lack of experience. The message these women get is, "We'll just wait until you have the baby," which is exactly the wrong message! Dentists and physicians should be encouraging pregnant women to go to the dentist as much as they can.

I know what you're thinking: "But, Doctor, how does bacteria up here in my mouth affect a growing baby down there?" There are two possible paths:

1. The oral bacteria or its toxins get into the mom's bloodstream, gaining access to the fetoplacental unit where they can cause a local infection or an inflammatory response.

2. The inflammation produced in the mouth in response to bacteria leads to an overall inflammatory response in the mother, and that process triggers negative outcomes for the fetus, like preterm birth.[12]

THE RISKS OF PRETERM DELIVERY

The last few uncomfortable weeks of pregnancy are a critically important time for a developing fetus. That's when the lungs and immune system mature. The baby prepares for a life outside its safe, sterile home, where all manner of environmental assaults from bacteria to viruses to toxins are just waiting to attack. Preterm birth is considered to be less than 37 weeks of gestation, and low birth weight is less than 5 pounds, 8 ounces. Globally, more than 15 million babies (one in every ten live births) are born too soon every year and, as a complication of preterm birth, more than one million children die annually. In the U.S., one in eight infants are born preterm. That's double the rate of other industrialized Western countries, like those in northern Europe.[13]

The earlier a baby is born, the more likely it is to have medical and/or oral health problems. Common challenges include difficulties with breathing, weight, body temperature, feeding, digestion, vision, hearing, and even some life-threatening conditions like brain or pulmonary hemorrhage, infections, or neonatal respiratory distress syndrome.[14] Even when babies recover from these conditions, many will continue to have significant related problems in the future, including poor motor skills, cognitive and intellectual impairment, and learning difficulties. They often have oral health problems, too, including enamel hypoplasia (thin enamel that makes teeth vulnerable to decay), dental crowding, and airway/sleep problems related to the underdevelopment of their jaws.[15]

For many years, we've known of risk factors that may affect or cause preterm birth, including smoking, consuming alcohol or drugs, poor prenatal care, poor maternal nutrition, and urinary tract infections. However, we have only recently learned about the significant impact of poor oral health and periodontal disease on the timing of birth.[16] As discussed earlier, the optimal time for periodontal intervention is before getting pregnant, but as the saying goes, "Better late than never." Recent research at the Perelman School of Medicine at the University of Pennsylvania found that women who were successfully treated for gum disease during pregnancy had only a 10.5 percent rate of premature birth as compared to 62 percent of the women where treatment was unsuccessful.

Periodontal disease has also been identified as a potential cause of low birth weight. Globally, 80 percent of newborns who die every year have low birth weight. According to the Centers for Disease Control and Prevention (CDC), "babies with a birthweight of less than 5.5 pounds may be at risk of long-term health problems such as delayed motor skills, social growth, or learning disabilities."

My aim here is not to add worry to parents who may already be worrying! I simply hope to raise awareness about the often-overlooked connections between maternal oral health and fetal well-being. It's not something doctors tend to mention, but oral care truly is an important part of prenatal care. As you've read before, the solution is relatively easy. All you need to do is go to the dentist, brush, floss, use safe and effective oral care products, and be aware of the condition of your teeth, gums, and airway.

DENTAL CARE DURING PREGNANCY

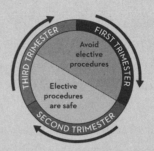

First Trimester: Traditionally, dental treatment, particularly elective dental procedures (root canals, extractions, restorations), are avoided during the first trimester. Emergency procedures, however, are typically indicated at all times during pregnancy. This is a good time to consult with your physician if you have any concerns.

Second Trimester: During the second trimester, the fetus undergoes tremendous growth and maturation. During this time, the teeth are forming and are susceptible to malformation. Even though you want to avoid extensive restorations and comprehensive treatment during pregnancy, necessary treatment is generally considered safer during the second trimester and the first half of the third trimester.

Third Trimester: After the first half of the third trimester, you should avoid elective procedures.

DENTAL X-RAYS

There are legitimate concerns about the use of X-rays during pregnancy, related to fetal exposure to ionizing radiation. To put things in perspective, though, a CT scan of the heart produces approximately 12 mSv of effective radiation, while two high-quality, intra-oral digital dental X-rays produce as little as 0.002 mSv.[17] In other words, you would have to take two high-quality/low-radiation dental X-rays every day for 16 years to equal the radiation of one CT scan.

The American Dental Association (ADA) and the U.S. Food and Drug Administration (FDA) agree on dental X-ray exposure during pregnancy. "Dental radiographs for pregnant patients may be prescribed according to the usual and customary selection criteria." Therefore, it is generally accepted that exposure to any dental X-rays required for the management of a pregnant patient in most situations should not place the fetus at increased risk. Having said all of that, if you're still concerned about X-rays during pregnancy (as a parent myself, I can certainly understand), there is good news: there are now no-radiation cavity detection technologies available! Ask your dentist.

OTHER MOUTH TIPS FOR PREGNANCY

Think about the mouths of all the caregivers in your child's life. The health of your mouth doesn't just matter prenatally for your baby. It affects your baby's ongoing wellness too! Early on, the bacteria in your mouth will seed your baby's bacteria. If you have "bad" bacteria in your mouth, you can pass it on to your baby. For example, transmission of Streptococcus mutans from mother to infant increases the risk of cavities. Did you think your child would inherit the color of your eyes and their dad's height but not the cavity-causing bacteria in your mouths? Think again!

Your new baby can catch cavity- and gum disease–causing bacteria from everyone around them, including parents, siblings, nannies, and grandparents. That even extends to the family pet! When I was in college, I came across a case where a child presented with advanced gum disease—way worse than you'd typically see in a child. They tested everyone in his family to try and find the source of the bacteria ravaging his mouth, but with no luck. Finally, in desperation, they tested the family dog and found the culprit. The dog was the source of the destructive bacteria that was making the child sick. Studies have actually shown that a dog or other pet with good oral health can help a child develop a healthy microbiome! This is not an argument for getting rid of your pet. However, you should be taking care of their mouth alongside everyone else's.

Wear your CPAP or oral devices if you have sleep apnea. This advice is good for anyone with sleep-disordered breathing, but especially so for pregnant women. As discussed earlier, pregnancy can make breathing more difficult and neither the pregnant mother nor the developing fetus can afford any interruption in oxygen.

CHEW ON THIS

If you had obstructive sleep apnea during pregnancy, you had a higher risk for a C-section. If you had a C-section, your baby may not have received the full spectrum of bacteria they would have gotten on the trip through the vaginal canal. A depleted microbiome can affect a baby's oral and physical health. In an effort to recolonize these infants, some physicians and midwives are taking swabs of vaginal bacteria from C-section moms and transferring them to the infant's skin. It's just one interesting example of modern medicine working with the complex biology that's been there all along.

The overarching lesson here: a parent's mouth can affect their baby's health in many direct and indirect ways. Taking care of your mouth is one very important way that you can take care of your baby.

THE BABY IS BORN!

The day has finally arrived. You're holding your precious baby in your arms, studying that little face, checking fingers and toes, and marveling at the magic of it all. How could you have made this fully formed human being? Maybe you're imagining everything you'll do together—the places you'll go, the adventures you'll have. It's all laid out before you like a tapestry. There's suddenly so much meaning in every little thing.

It's amazing how many feelings can happen at the same time. While you wonder at the miraculous nature of life, you're already worrying about all the decisions, the responsibility, and the immense weight of ushering a new life through the world. While you cry with joy, fear is there, too, along with worry and love and sadness and physical discomfort and the most profound sense of peace and contentment you've ever felt. Welcome to the happy/sad, thrilling/terrifying, beautiful/ugly life of a parent!

One of the first things I learned in my parenting journey was that decisions are hard. Some can come and go with little consequence, while others can change the direction of your baby's life. Sometimes, especially in those first few weeks, it feels impossible to know the difference. Some of the earliest and most meaningful decisions parents make are related to a baby's oral health and development. And one of the most critical of all is whether or not a mother decides to breastfeed. There are many reasons why a new mother might not be able to breastfeed. It's a complicated decision. But if you can do it—even for a limited time—the benefits to your baby will be profound.

Minutes after a healthy baby is born, it starts to hunt for the breast. Food is the first thing on a baby's mind, and rightly so. Colostrum, that first milk, is so full of life-giving ingredients, it's like gold to a newborn. It contains antibodies, body-building fats, and just the right blend of sugars to feed a brand-new microbiome seeded by the mom's skin and by the birth itself. Colostrum and the breast milk that will follow in the coming days truly are the perfect food for an infant. And not just any infant! Each mother's body creates the perfect blend for her individual baby. It's a match made in . . . evolution.

After lactose and fats, the third-most-abundant ingredient in breast milk is a group of complex sugars called human milk oligosaccharides, or HMOs. Strangely, they cannot be digested by the newborn child.[18] For many years, scientists knew these HMOs had to have some purpose, and now we know: it turns out they are food for the baby's microbes! HMOs are an example of the prebiotics I mentioned in the last chapter. They create the perfect, food-filled environment for the right kind of healthy mouth and gut bacteria to thrive. In addition to HMOs, mother's milk also introduces live microbes that help colonize the baby's developing digestive and immune systems.[19]

In addition to the digestive, nutritional, and immune-system benefits of breastfeeding, the shape of a breast as it flattens against a baby's palate helps that palate develop into the proper shape. The sucking motion strengthens the baby's cheeks and jaw, tones the tongue so that it rests on the palate in the proper way, and teaches the baby to breathe through the nose. This helps the jaws grow correctly, reducing future dental crowding.[20] This also helps the airways develop properly, preventing breathing problems like sleep apnea. Properly developed face and mandible muscles allow for the anterior positioning of the mandible, leaving room for the tongue, again resulting in less risk of snoring and apnea. All of this can reduce colic and reflux, which can cause enlargement of the tonsils and adenoids (and affect breathing).

The American Academy of Pediatrics (AAP) issued its first document in support of breastfeeding in 1948 and has since updated its support several times with stronger statements as we continue to learn more and more about breastfeeding's benefits. The current summary page of AAP relating to breastfeeding states: "Breastfeeding is a natural and beneficial source of nutrition and provides the

healthiest start for an infant. In addition to the nutritional benefits, breastfeeding promotes a unique and emotional connection
between mother and baby."

In the policy statement "Breastfeeding and the Use of Human
Milk," published in the March 2012 issue of *Pediatrics*, the AAP reaffirms its recommendation of exclusive breastfeeding for about the
first six months of a baby's life, followed by breastfeeding in combination with the introduction of complementary foods until at
least 12 months of age, and continuation of breastfeeding for as
long as is mutually desired by mother and baby.[21]

Breastfeeding provides a protective effect against respiratory
illnesses, ear infections, gastrointestinal diseases, and allergies,
including asthma, eczema, and atopic dermatitis. The rate of sudden infant death syndrome (SIDS) is reduced by a third in breastfed
babies, and there is a 15–30 percent reduction in adolescent and
adult obesity in breastfed infants. "As such, choosing to breastfeed
should be considered an investment in the short- and long-term
health of the infant, rather than a lifestyle choice."[22]

Breastfeeding doesn't just benefit the baby. The mom benefits,
too, in some very significant ways. Nursing reduces postpartum
bleeding and hemorrhage risk because it increases the production
of oxytocin, which helps the uterus contract.[23] That same oxytocin is responsible for the joyful feeling mothers experience during
breastfeeding. Each additional year of breastfeeding has been
shown to reduce breast cancer risk by 4.3 percent.[24] Each month
of breastfeeding reduces ovarian cancer risk by 2 percent and has
been shown to be lower overall in those who have ever breastfed.[25] Mothers who breastfeed longer may also be at a lower risk
of developing multiple sclerosis.[26] And nursing moms can generally return to their pre-pregnancy weight more quickly than those
who don't.

Breastfeeding—and difficulties with it—is the perfect example of how our fragmented health care system can disadvantage our children from the minute they're born. Each care provider is limited to his or her specialty, and few manage to consider the full picture. Often, new moms aren't supported enough in their breastfeeding journey. Sometimes milk takes a day or two longer to come in, and well-meaning obstetricians, concerned about weight gain, push formula. Often, they just needed to wait one more day, but introducing formula so early makes establishing breastfeeding that much harder. Latch issues are sometimes too painful and women lack the support of a lactation consultant or caring midwife who could help. Or, there is a physical problem with the baby, like a tongue- or lip-tie, that doctors miss or don't take seriously.

When my son was born, some of the nurses and doctors were immediately ready to give up on my wife's breastfeeding. We insisted that we would breastfeed, but even then, we struggled. The medical establishment does a huge disservice to women when it fails to tell them that breastfeeding can be difficult. It can! It's hard! It doesn't always come naturally. But with proper information and support, most women can achieve a wonderful, healthy, bonding breastfeeding relationship with their babies.

However, if you can't breastfeed—or didn't breastfeed—don't worry. There are plenty of other things you can do to support your baby's oral health, and you will learn them all in this book.

INFANT ORAL CARE ISSUES

While breastfeeding is an excellent way to prevent many common oral health problems in a growing baby, problems still do occur.

Here is a quick primer on some of the infant oral health issues I commonly see, and some tips for how to spot them early.

THE TONGUE

In relation to its size, the tongue is the strongest muscle in the body (technically it is a group of muscles). It's also one of the most important since it is involved in many aspects of life, including sucking, swallowing, taste, jaw development, airway development, and speech.

Poor tongue habits can cause all sorts of issues with breast-feeding, speech, swallowing, snoring, sleep apnea, orthodontic malocclusion (a misalignment of the teeth and bite), and much more. Some babies develop tongue habits that may interfere with a good latch during breastfeeding. For example, babies sometimes suck their tongues, hold them in the back of their mouths, or use them to push the nipple out of their mouths.[27] Habits like these can make it difficult to establish or maintain breastfeeding, but they are habits that can be changed. An experienced lactation consultant can often help.

Another tongue-related habit that can become problematic is tongue thrust. This happens when the tongue protrudes through the front teeth during swallowing or speech and while the tongue is at rest and in a relaxed position. Factors that can contribute to tongue thrusting include tongue-tie, an enlarged tongue, thumb-sucking, the use of bottles or pacifiers, airway obstruction, or hereditary factors. Since we swallow up to 2,000 times a day, the pressure of the tongue against the teeth during swallow-ing in someone with tongue thrust can force the teeth out of alignment and cause spacing and other orthodontic issues.

Tongue thrusting can also lead to poor jaw development, periodontal problems, airway problems, and speech difficulties. I always recommend treating tongue thrust as early as possible (as I do for many pediatric oral-health problems).

Ankyloglossia

Try saying that three times fast. No wonder it means "tongue-tie"! Tongue-tie happens when the thin membrane under the tongue (the lingual frenum) is too tight to permit the tongue to move freely in the mouth. According to current literature, tongue-tie occurs in up to 10 percent of newborns, but I wouldn't be surprised if the true number is significantly higher.[28] Many cases go undiagnosed or unreported. A baby with tongue-tie cannot extend his tongue forward far enough to help create a good seal during breastfeeding. The tongue doesn't rest correctly on the palate so the palate may not grow correctly, leading to smaller arches, a higher palate, a smaller airway, smaller jaws, dental crowding, and reflux, which leads to colic and inflamed tonsils or adenoids, which also make breathing more difficult. Tongue-tie can contribute to reduced oxygen flow to the body, which can have a significant impact during the early weeks of the infant's brain and nervous system development. Clearly, this is an important one to rule out early. Sometimes tongue-ties can be difficult to diagnose. Here is a helpful checklist of common signs and symptoms that should inspire a trip to the pediatric dentist.

SIGNS OF A TONGUE-TIE

- Poor latch, milk drooling out
- Baby arching away from the breast in frustration
- Pain and injury to the nipple from the baby clamping the gums together in frustration
- Fussy baby during feeding (a baby who is getting enough milk is usually relaxed and happy)
- Reflux
- Vomiting after breastfeeding
- Failure to thrive or slow weight gain
- Infection in the breast due to inadequate emptying during feedings (mastitis)
- Cracks in nipples that won't heal
- Clicking and swallowing air when latched (aerophagia)
- Falling asleep during nursing (though many babies without tongue-tie do this too)
- Baby swallowing air (because of a poor latch)

Revising or releasing tongue-ties used to be a common procedure for family physicians, pediatricians, or midwives, but unfortunately this has fallen out of practice. Long ago, midwives would cut the tongue-tie with their fingernail (they kept one nail long and sharp for this purpose), as the survival of the baby was dependent on their ability to feed.[29] Of course, as we learned more about infections, this method was replaced with other techniques. Today, there are many competent dentists and physicians using a variety of tools to release or revise tongue- or lip-ties, but the state-of-the-art treatment of choice is a soft-tissue dental laser. These are safe, quick, and effective, don't require any type of numbing, and heal quickly. This is so easy; I'm always perplexed as to why it's not done routinely. Doctors will often wait and see, but that approach doesn't make much sense in this case.

I will never forget the first time I met Chloe. She was only two-and-a-half years old when she came along with her older brother for their first pediatric dental exam with my wife. She called me over to do an orthodontic screening on the older brother, but Chloe's bigger-than-life personality had filled the room with laughter and excitement. She had curly red hair and big green eyes, and she was making all sorts of funny faces, entertaining her brother and our dental crew. Once I stopped laughing, I asked her how old she was, and she raised two of her fingers, indicating her age. Immediately afterward, her mother told us Chloe wasn't talking yet, and even though she was quite concerned about that, her previous dentist and pediatrician had suggested waiting to see if she improved. My wife and I examined Chloe's mouth and the first thing that jumped out at us was a tongue-tie. With the mom's permission, we released the restriction of Chloe's tongue in just a few minutes. When the family returned for their periodic follow-up visit six months later, the mom grabbed my wife's hands and started kissing them. She thanked us from the bottom of her heart. Chloe was talking! And, according to her mom, she was "making up for lost time."

Tongue-ties don't tend to self-correct. And when they can cause so many problems, why not just treat them right away? The longer you wait, the more likely you are to have secondary problems related to the tongue-tie. For example, after age two, releasing a tongue-tie often goes hand in hand with speech therapy. The tongue doesn't just find the correct resting position on its own when you wait that long.

Unfortunately, misdiagnosis is a common occurrence in cases like Chloe's. Just recently, I read about a six-year-old boy in Texas who could barely speak.[30] He could pronounce the beginning of words, but not the end of them. He was speaking at about the level of a one-year-old. His parents believed he was nonverbal because of a brain aneurysm he had when he was only 10 days old, and they'd had him in speech therapy since he was about one. According to his mom, there were other issues too: "Sleeping was always stressful. He would stop breathing. He had trouble eating and swallowing; every single meal we would have to remove something that was choking him. He didn't get the nutrition he needed. His teeth started having problems."[31]

After years of frustration, a routine dental visit at the age of six revealed the little boy's tongue-tie. It took the new pediatric dentist about 10 seconds to revise the tie. Later that same day, the little boy said, in a voice as clear as day: "I'm hungry. I'm thirsty. Can we watch a movie?" His parents were astounded! Suddenly, their son could speak!

"It's like night and day. He doesn't have choking episodes anymore; he's eating different types of food," his mother said. "He's behaving much better at school. His behavior was a problem because he was getting poor quality of sleep at night, he was constantly tired, and was not able to express himself. He doesn't snore anymore. He doesn't have sleep apnea anymore."

It's a heartwarming story, but heartbreaking too. This little boy could have been spared years of speech, sleep, school, behavior, and developmental challenges with a simple procedure performed just after birth.

Similar to a tongue-tie, with a lip-tie, the lip frenum that attaches the lip to the gums can be too thick or tight. It can also cause problems with breastfeeding, cavities, lips that can't close fully at rest, poor breathing habits, or a space between the front teeth. In a recent study, researchers found that 84 percent of infants who had died from SIDS had apparent restriction of the upper lip, as opposed to only 5 percent in the general population.[32]

More reasons to treat tongue- and lip-ties early:

- It makes speaking easier! In some cases, it makes speaking *possible*.
- It allows a person to lick.
- It helps a baby to breathe better.[33]
- It can avoid the frenum pulling on the gingiva, leading to gum recession.
- It can alleviate gagging while feeding.
- It can stop sleep disorders ranging from snoring to sleep apnea.[34, 35]

NATAL AND NEONATAL TEETH
Natal teeth are teeth that have already erupted into the mouth at birth, and neonatal teeth are teeth that emerge through the gum during the first month of life (the neonatal period). Although it's possible that these could be supernumerary teeth (extra teeth), they're commonly the child's normal baby teeth that have erupted prematurely. You need to see your pediatric dentist for this since possible intervention and/or treatment may vary depending on the situation.

TEETHING
Although the age when babies get their teeth varies, teething ordinarily begins around six months of age and may continue until the child is about two years old. The order in which the teeth erupt

is almost always the same. The lower front teeth usually come in first. Some babies may not get their first teeth until 12 months or even later. However, if the teeth are significantly delayed, especially if a tooth is delayed on one side compared with its contralateral side, you may want to have a pediatric dentist evaluate the situation to make sure the baby is not missing any teeth, or the tooth is not blocked.

If there is discomfort or fussiness during teething, I would not recommend any medications, including oral gels. You can use safe teething rings that have been refrigerated (never frozen, since this can be too cold for the baby's mouth). Just remember to be aware of the quality and source of anything your baby puts in their mouth. Recently, a baby ended up with severe levels of lead poisoning from a teething bracelet the parents had purchased from a local fair.[36] Teething jewelry can have other risks, too, including injury or an infection in the mouth, or worse!

In 2018, the U.S. Food and Drug Administration (FDA) had to issue a warning to parents after an 18-month-old died, strangled by his amber teething necklace during a nap, and a seven-month-old choked on the beads of a wooden teething bracelet and was taken to the hospital. "We know that teething necklaces and jewelry products have become increasingly popular among parents and caregivers who want to provide relief for children's teething pain, and sensory stimulation for children with special needs. We're concerned about the risks we've observed with these products and want parents to be aware that teething jewelry puts children, including those with special needs, at risk of serious injury and death," the FDA said in its statement.[37]

Babies can be irritable and fussy during teething, but it doesn't cause illness. So, if your child has prolonged crying, earache, diarrhea, fever, coughs, or refuses to feed, please take them to your doctor for a checkup.

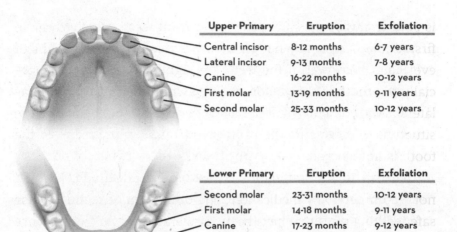

Upper Primary	Eruption	Exfoliation
Central incisor	8-12 months	6-7 years
Lateral incisor	9-13 months	7-8 years
Canine	16-22 months	10-12 years
First molar	13-19 months	9-11 years
Second molar	25-33 months	10-12 years

Lower Primary	Eruption	Exfoliation
Second molar	23-31 months	10-12 years
First molar	14-18 months	9-11 years
Canine	17-23 months	9-12 years
Lateral incisor	10-16 months	7-8 years
Central incisor	6-10 months	6-7 years

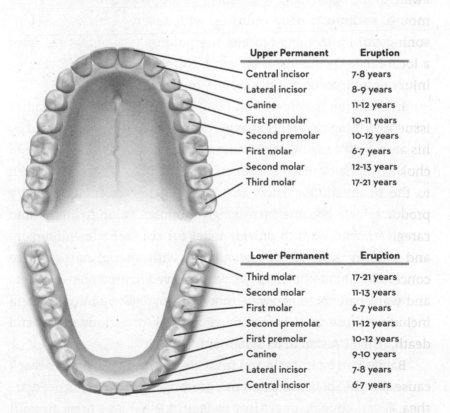

Upper Permanent	Eruption
Central incisor	7-8 years
Lateral incisor	8-9 years
Canine	11-12 years
First premolar	10-11 years
Second premolar	10-12 years
First molar	6-7 years
Second molar	12-13 years
Third molar	17-21 years

Lower Permanent	Eruption
Third molar	17-21 years
Second molar	11-13 years
First molar	6-7 years
Second premolar	11-12 years
First premolar	10-12 years
Canine	9-10 years
Lateral incisor	7-8 years
Central incisor	6-7 years

THUMB-SUCKING

Thumb-sucking is a behavior found in humans and some other primates. It can start even before birth, as early as 15 weeks from conception, and can continue well into childhood, even adulthood, if not managed correctly. The longer it continues, the greater effect it can have on the growth of the teeth, jaws, bite, and even the face itself. When you suck your thumb, the pressure of buccinator (cheek) muscles on the sides combined with the force of the thumb on the front teeth, constricts the posterior teeth, causing a crossbite, pulls the upper jaw forward, and lifts the front teeth open. The recommended age to stop thumb-sucking can vary with every child and needs to be discussed with your doctors. However, children who suck their thumb after age four are three times more likely to develop a crossbite than those who stop at age one.[38]

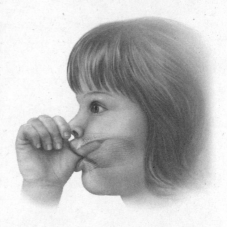

As you might expect, I always try to get patients to stop these behaviors early (the younger the child, the easier it is to break a bad habit). I once had to treat a mother for thumb-sucking. She was worried that it was affecting her driving. Treatment options get more invasive as patients get older. For a young child, treatment could be as easy as wearing a sock over the hand at night and/or wearing a band on the elbow for a couple of weeks. To treat older children, we may use an orthodontic device to physically block the thumb from entering the mouth. For teenagers, orthodontics, including a jaw expander, may be necessary. Adults often require a combination of orthodontics and jaw surgery. In addition to the pleasurable and soothing experiences that lead to continued thumb- or finger-sucking, the action of raising the arm and putting it in your mouth can be addictive on its own. Just think of how many times we all now raise our arms to look at our mobile phones.

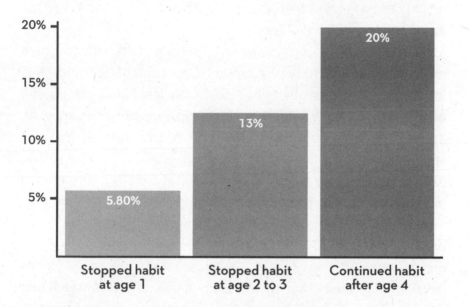

Thumb-Sucking or Pacifier Use and Incidence of Crossbite

BABY BOTTLE TOOTH DECAY

Once your baby's teeth have come in, letting your child fall asleep during feeding can cause major dental problems. Seventeen percent of parents put their child to bed every night with a bottle of milk, formula, or juice. When your baby is sleeping, the milk or juice pools on the teeth.[39] The enamel on baby teeth is thinner than on permanent teeth, and the pulp is relatively larger, so dental decay can spread to the nerve faster. The treatment for severe cases includes pulpotomies (removing the infected nerve), crowns, or extraction of the affected teeth—usually the front baby teeth. Baby teeth are critical for a variety of reasons, including keeping the needed space for permanent teeth. Losing them at such a young age can create a nightmare scenario for both parents and the pediatric dentist, not to mention how devastating it could be for the child. Wiping your child's gums and teeth with wet cotton gauze after feeding will remove the sugar and bacteria that could cause decay. It's important to note here that although breast milk is better (and has less sugar) than formula, either one can cause the development of this problem.

Once your baby is physically ready to sleep through the night without being fed (typically around 5 months of age and/or 15 pounds, which is usually right before baby teeth start coming in), you can wean your baby from night feeding using a variety of techniques. Do some research and choose whatever method you are comfortable with. As an added bonus in addition to protecting their teeth, your little one will be more likely to sleep through the night if the sleep association with feeding is gone.

USE OF PACIFIER OR BOTTLE

Even though there are conflicting data on some potential benefits versus dangers of using pacifiers, my personal recommendation is to avoid pacifiers if possible. If you must use one, try to avoid using it past six months so it doesn't interfere with the growth of the mouth. The breast shapes itself to a baby's mouth but pacifiers and bottles make mouths form to *them*. If you are bottle-feeding, then you and your pediatric dentist need to monitor your baby's mouth closely.

A NOTE ABOUT BABY TEETH

Healthy baby teeth are so important. The 20 baby teeth (primary teeth) that typically erupt between the ages of six months and two years are as important for physical health as they are for the psychological well-being of the child. Baby teeth are more than just a practice round for adult teeth—they're perfectly suited to the mouths of babies: the right shape and size to fit small mouths, to manage first foods, to help form first words, to guide the growth of the jaws and the face, and to hold space for the bigger teeth to come.

Baby teeth make their appearance when we are first learning to talk, to relate to others, and make sense of the impact of our actions and words. Losing baby teeth too early can impede that critical process. If baby teeth fall out too early, permanent teeth are likely to come in crooked and crowded, or they'll get impacted and won't come in at all. And baby teeth help create beautiful smiles that give children their confidence and melt our hearts. Baby teeth, just like babies themselves, are precious.

HEALTHY, CRUNCHY, HARD FOODS

Once your child is a little older and there is no choking hazard, make sure you use healthy, raw, hard, natural food as much as possible, not only to provide the nourishment needed for the developing mouth and teeth, but also to continue the correct usage of the mouth and facial muscles. Think fresh carrot/celery/apple (hard, healthy) versus fruit juice or applesauce (soft, full of sugar). When it comes to the mouth, if you don't use it, you lose it!

WHEN TO SEE THE PEDIATRIC DENTIST

One thing I've learned over many years and tens of thousands of patients is that there is a ton of confusion about when to see a pediatric dentist. The American Academy of Pediatric Dentistry states: "In order to *prevent dental problems*, your child should see a pediatric dentist when the first tooth appears, or no later than his or her first birthday." This is much earlier than many doctors recommend—another example of the disconnect between dentistry and medicine. In fact, only 16 percent of children see a dentist by their first birthday, and 20 percent of them don't see a dentist until they are five or older.[40]

As you are already learning, a dentist can and should go further—preventing more than just *dental* problems—so I recommend a consultation with your pediatric dentist even before your baby is born. This will help you to be proactive about the things you can do to set your baby up for a lifetime of good habits and a healthy mouth. Before your baby is born is the perfect time to set up a dental home for your family. Ideally, your pediatric dentist will be in the same office as your own dentist so they can collaborate and deliver coordinated care. If you've already had your baby, that's okay! It's never too late to establish a dental home.

PARENTS' CHECKLIST

Pre-Pregnancy
- Establish a dental home for you and your family.
- Future dads: Go to the dentist! Poor oral health can reduce sperm count and quality.
- Future moms: Go to the dentist! Poor oral health can negatively affect your fertility.

Prenatal
- **Diet:**
  - Reduce your consumption of sugar and processed foods.
  - Eat foods with plenty of vitamins K2 and D.

- Consider taking prenatal supplements that include vitamins D3 and K2; the majority of women are significantly deficient in these vitamins (consult with your doctor, of course, before changing your dietary regimen).[41]

- Maintain regular visits to the dentist, especially during the second trimester and the first half of the third trimester.

- Take great care of your mouth at home. Great oral health helps support a healthy pregnancy and also ensures a healthy microbiome is passed on to the baby post-birth.

- Only use safe and effective oral care products (see Chapter 5).

- Take care of the mouths of all future caregivers. Encourage all family members and caregivers to see a dentist and make oral health a priority.

- Take care of your airway health—wear your CPAP or other oral devices as indicated.

- Consider natural (vaginal) birth.

Birth–6 Months

- Consider breastfeeding, even if it's partial or for a limited time.

- Continue with your own healthy diet, especially if you're breastfeeding.

- Resolve any oral issues early. See a lactation consultant, revise tongue- or lip-ties.

- Commence your baby's pediatric dental visits and continue visiting your own dentist.

- Work with your pediatric dentist to address any poor oral habits or other issues.

- Wean your baby from night feedings by six months.

- Don't let your baby fall asleep during feedings once baby teeth start erupting.

6–24 Months

This is a key period for a child's oral health. This is when all the baby teeth erupt, so it's critical to start things off right. The oral microbiome is evolving and maturing, you'll begin to introduce oral care products, and your child will begin to build psychological associations with their mouth and oral care, whether positive or negative.

During this time, it is crucial to:

- Continue visiting your pediatric dentist (and your own) every three to six months.
- Establish a nutritious and varied diet to nourish not only your baby's growing body, but also their rapidly developing mouth.
- Introduce a small toothbrush with ultra-soft bristles right as baby teeth are coming in (babies love getting their gums massaged when teething). Initially, just use the toothbrush without any toothpaste until your child is comfortable with the brushing routine.
- Once your baby is accustomed to the act of brushing, introduce toothpaste, but not any old toothpaste. Due to potential safety concerns regarding excess fluoride and young children, I would use a toothpaste with hydroxyapatite instead. You can find a comprehensive list of beneficial (and harmful) ingredients in Chapter 5.
- Consult with your pediatrician to make sure your child is getting adequate amounts of vitamins D and K2 either through diet, oral care products, or supplements.
- Introduce a safe and effective floss once there are contacts between the teeth. Avoid floss that contains PTFE (polytetrafluoroethylene), petroleum, or other potentially toxic materials—find more specifics in Chapter 5.

2–6 Years

Did you know that five-year-old kids brush only 25 percent of their teeth? Even 11-year-olds brush only 50 percent of them. Developing good brushing habits early is key for keeping your child's mouth healthy as they grow. Here's how:

- Start with brushing for them and then gradually supervise until they learn to brush on their own. This may take years, so be patient. Also, make brushing fun! Brush together, make it a game, do a dance, whatever it takes to get them brushing. Children brush 73 percent longer with music, so get a musical toothbrush.
- Don't forget to brush the tongue or use a gentle tongue scraper.
- Change your child's toothbrush (and your own) every one to three months depending on the type of the bristles. Some of the softer/better bristles need to be switched more frequently since they lose their effectiveness quicker.
- Continue to use a toothpaste with hydroxyapatite and/or fluoride indefinitely. (See Chapter 5 for more info on this.)
- Floss at least once a day indefinitely.
- Maintain routine visits to the pediatric dentist indefinitely.

6+ Years

Kids start losing their baby teeth around age six, and they get most of their permanent teeth between the ages of six and twelve. This is a critical time for the development of the mouth.

- Keep a close eye on the permanent first molars (aka six-year-old molars). These are the most important molars and they're sometimes neglected because they erupt while the child still has most of their baby teeth.

- Either manual or electric toothbrushes are fine, as long as they are high quality; your child should use whichever they prefer. Change the brush every one to three months.
- Introduce a gentle tongue scraper if you haven't yet. The tongue accumulates lots of microbes and is a common source of bad breath.
- Introduce a safe and effective pH balancing mouthwash, ideally containing hydroxyapatite and prebiotics—as soon as your child learns how to swish and spit. Avoid any mouthwash with alcohol, artificial colors or flavors, antibacterial ingredients, or acidic pH (you can check the pH with strips or a meter). Many oral care products are extremely acidic because of their ingredients, or in order to extend shelf life. Fluoride mouthwash should never be used before age six because children may swallow it. (See Chapter 5.)
- Introduce a safe and effective pH balancing mouth-spray. Children should use the spray after every food or drink to alkalize the pH of their saliva. (See Chapter 5.)
- Visit an orthodontist no later than age seven to evaluate the airway, crowding, bite, skeletal growth, habits, tooth eruption, and so on.

CONNECTING THE DOTS

The mouth is one of the most critical organs for your child's health and development. You can ensure its proper growth by keeping your own mouth in top shape, and by addressing any issues that come up early. When it comes to the mouth, "wait and see" is rarely the answer.

CHAPTER 3

BREATHING, SLEEP, AND YOUR MOUTH

IN THIS CHAPTER

- Why sleep is important.
- How your mouth affects your sleep.
- What happens to your body when you don't sleep well (or enough).
- The connection between sleep-disordered breathing and overall health.

A deep, restorative sleep makes everything feel possible. It clears the mind and heals the body. Some people sleep well every night with little problem. But for most of us, a fantastic sleep is a rare unicorn. Most nights, we sleep poorly. When we're awake, we live on caffeine. We're irritable, moody, overwhelmed, and frustrated, and struggle to get through the day. It's common to blame poor sleep on kids or bodily discomforts, noisy bedmates, or stress. But in many cases, the culprit is more insidious. The real sleep thief isn't environment or anxiety, kids or pain—it's how you breathe.

It goes without saying that breathing is important, but I'm going to say it anyway. We need oxygen every second and can survive for only a few minutes without it. We breathe 12–16 times per minute, taking upwards of 20,000 breaths per day. Every cell in the human body needs oxygen, and without it, the body panics and goes on high alert.

Yet half of us suffer from airway obstruction that interrupts our sleep.[1] When breathing stops during sleep, the body releases a burst of adrenaline, the heart rate spikes, and the brain wakes the sleeper up. The person can breathe again, but now has adrenaline coursing through their body. This is hard on organs, especially the heart, and, of course, adrenaline makes getting back to sleep more difficult.

Sleep may seem strange and even unnecessary if you look at it superficially: we close our eyes, lie down somewhere, and disconnect from the world for several hours. In today's busy world, it often feels like an annoyance. Maybe that's one reason why the majority of people—a whopping 79 percent—don't get enough.[2]

The quality of your life depends upon the quality of your sleep, and your mouth plays a critical role in that process. Poor sleep can impact your stress hormones and thyroid, distress your immune system, and affect your mood, brain function, and metabolism.

CHEW ON THIS

Study after study has revealed that seven to eight hours is the ideal sleep duration for most people.[3] However, there are exceptions.

Albert Einstein famously slept from 10–11 hours a night to "nurture the creative process," as he put it, which seems to have worked out pretty well for him.[4]

On the other hand, Thomas Edison, the inventor of the light bulb (and, as the creator of artificial lighting, the man often blamed for much of humanity's sleep deprivation), slept only three to four hours a night. He considered sleep a waste of time and "a heritage from our cave days."[5]

BREATHING AND YOUR MOUTH

In Chapter 1, we learned about the mouth's anatomy, but here's a quick review: both the maxilla and mandible are involved in forming the airways. The maxilla forms the bones of the nasal cavity and shapes the palate, which is both the roof of the mouth and the floor of the nose. The mandible houses the tongue, which is made up of several muscles. In the front of the mouth, the tether holding down the front of the tongue is called the frenulum and it's usually free to move around unless there is a tongue-tie (see Chapter 2). In the back of the mouth, the tongue is anchored to the hyoid bone.

When relaxed, the tongue is supposed to gently rest on the roof of the mouth against the palate, right behind the teeth. The teeth rest slightly apart, and the lips stay gently closed. Breathing happens through the nose almost all the time. In its ideal position, the tongue helps with breathing by aiding with normal growth of the

airways, and also by holding the airways
open. But when the tongue cannot sit in
its ideal position due to a narrow palate,
small jaws, mouth-breathing, poor tone,
or a tongue-tie, it may move backward
and block the airway, interfering with
breathing and impacting oxygen intake.[6]

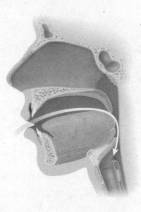

This often gets worse during sleep,
when muscles relax.

WHY IT'S CRUCIAL TO BREATHE THROUGH YOUR NOSE

Unless you're exercising hard or have temporary congestion, you
should be breathing through your nose. The nose, mouth, and
pharynx act as tubes for air to travel through. They also humidify
and warm the air going in. Your lungs need the air coming in to be
within a few degrees of your body temperature, and nearly satu-
rated with humidity.[7]

The nose has evolved to be extremely good at exchanging
moisture and heat. It's far better at this than the mouth. Breathing through the nose also slows down airflow, allowing it to mix
with the nitric oxide released in the maxillary sinuses.[8] It has been
shown that blood is oxygenated 10–15 percent more when you
breathe through your nose. Nitric oxide is a very important molecule in the body that helps in many physiological processes such
as expanding the blood vessels and increasing blood flow. In fact,
it has been called the "miracle molecule."[9]

HOW NOSE-BREATHING KEEPS YOU HEALTHY

- It removes unwanted particles such as dust and microbes from the air before they enter the lungs.
- It helps you breathe deeper, drives oxygen more efficiently to the lungs, and keeps CO_2 levels in a healthy range.
- It helps activate more of the parasympathetic nerve receptors involved in calming and reducing stress.
- It can lower heart rate and blood pressure.
- Air resistance and stimulation from nose-breathing help the upper jaw grow correctly.
- It helps position the tongue against the palate for the healthy growth of the mouth and airways.

THE NEGATIVE IMPACT OF LONG-TERM MOUTH-BREATHING

Chronic mouth-breathing can lead to many harmful outcomes that can compromise your quality of life, such as changes in body posture, oral health ramifications, and sleep disorders.

Children with obstructed airways (narrow airways, chronic congestion, enlarged adenoids, or a deviated septum) who are mouth-breathers do not experience the normal air resistance that stimulates healthy nasal and oral growth. As a result, they tend to have smaller mid-faces, nasal cavities, and upper jaws (maxilla), and narrower and higher vaulted palates that encroach into the nasal spaces.[10] As we've discussed, breathing through the mouth keeps the tongue low, which prevents it from helping to shape the palate and align the teeth. The tongue is nature's palatal expander—having it in the wrong position can cause dental crowding, impacted wisdom teeth, and airway obstruction.

The growth of teeth depends, in part, on their contact with each other. In a mouth-breather, the top and bottom teeth don't touch. 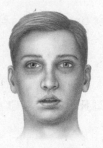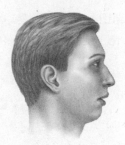 This "tells" the teeth to keep growing. Sometimes this just causes crowding, but it can also cause a gummy smile, or a skeletal open bite known (aptly) as "long face syndrome."

Mouth-breathing can also lead to dry mouth, which can have its own detrimental effects on oral health, causing cavities, bad breath, and gum disease. It can also cause heartburn and throat and ear infections. An obstructed airway in a mouth-breather can even lead to postural problems and slouching![11]

It is not common knowledge among most parents that mouth-breathing is a cause for concern. Many doctors will even underplay or dismiss it, assuming it's due to congestion, or that the child will grow out of it. But, unless they are sick, children should be breathing through their noses. Consistent mouth-breathing in a child may be a sign of a problem that needs to be addressed. An obstruction that *isn't* addressed, like enlarged adenoids, can affect the child's physical and intellectual development.

We take our breathing for granted, and we turn snoring into a joke. But if you're not breathing well, your cells are in trouble. Your body and mind are literally suffocating.

The good news? (It's definitely time for some!) There are many treatments for airway problems: mouth, tongue, and jaw exercises, dietary changes, devices to help you breathe when you sleep, orthodontics, surgeries, and more, and I discuss some of these throughout this chapter. It's worth investigating your options because, as with anything, different solutions work for different people. But one thing is true for everyone: we all need to breathe!

TRAIN YOURSELF TO BREATHE BETTER

The diaphragm is a muscle at the base of the lungs. When you breathe in, your diaphragm contracts to allow more air into the lungs. When you exhale, it relaxes. Diaphragmatic breathing encourages higher oxygen intake, slows down the heartbeat, and may help you calm down. We all know innately how to breathe deeply using our diaphragms, but as we get older, we tend to forget. The constant stresses and worries of life get us into the habit of breathing more shallowly, using our chests. If you practice regularly, you can train your body to breathe properly on its own.

Try these diaphragmatic breathing exercises for just three to five minutes per day:

- Sit comfortably in a chair with your back straight. Put one hand on your belly and the other on your chest.
- Relax your muscles: start with your toes, then legs, back, belly, shoulders, neck, and all of your facial and mouth muscles.
- Inhale deeply and slowly through the nose (for three to five seconds), letting your belly rise, while the chest remains still.
- Exhale slowly through the nose, while gently tightening your abdominal muscles. The hand on your belly should feel the air moving out.
- Pause for a couple of seconds, then repeat 12–15 times.

TONING YOUR TONGUE

Your tongue is a muscle. Like other muscles in your body, it can lose its tone and position if it's not used correctly. A flaccid tongue can interfere with breathing, especially during sleep when muscles are relaxed. Here are some exercises to strengthen your tongue muscles:

- While your mouth is open, suck your tongue upward against the palate so that the entire tongue is against the roof of your mouth—repeat 20 times.
- With your mouth open, push the tip of your tongue up against the front part of your palate and slide it backward—repeat 20 times.
- Force the sides of your tongue downward against the floor of your mouth while keeping the tip of your tongue in contact with the lower front teeth—repeat 20 times.

WHY DO WE NEED SLEEP?

Sleep is restoration, recovery, and rebuilding. It heals injury, stores memory, and refreshes the brain. Amazingly, despite many decades of intensive research, there are elements of sleep that still remain a mystery. We don't exactly know how dreaming benefits memory, for example, only that it seems to do so. But we do know for certain that adequate sleep is essential for good health.

People can be very sensitive to even moderate sleep inter-ruptions. This is especially true for kids. When children don't get enough sleep, they may become irritable, impulsive, overly sensitive, and difficult to communicate with. This is also the case for many adults.

Sleep deprivation can lead to fatigue, chronic pain, obesity, behavioral issues, concentration and performance challenges, anxiety, depression, and a variety of systemic diseases. It can also raise a person's risk of developing dementia and Alzheimer's. Correcting breathing issues is one of the most meaningful things a person can do to improve their sleep, which leads to an improvement of quality of life, relationships, and long-term health prospects. Healthy breathing starts with the healthy growth and development of the mouth. As with other mouth-related problems, earlier treatment is easier and less costly. But breathing problems can be treated at any age.

As we've discussed in previous chapters, genetic and evolutionary factors have left humans with smaller mouths. Epigenetic changes related to the modern diet, breastfeeding, and environmental toxins have also affected the mouth and airways. Then there are common problems like a deviated septum or narrow nasal passages, poor tongue posture, mouth-breathing, an increase in allergies and asthma, and poor oral habits, such as thumb-sucking or pacifier use. Finally, dentists play a role.

By focusing only on fixing cavities and straightening crookedness, dentists may fail to see the mouth for anything more than just a collection of teeth. But early, timely intervention by an astute dentist or orthodontist can help with healthy growth and development, creating more room for permanent teeth (and less need to extract teeth), a better bite/occlusion, a more proportional face, and a better airway, resulting in better sleep.

WHAT HAPPENS DURING SLEEP?

During sleep, the nervous system is inactive, voluntary muscles are inactive and relaxed, and consciousness is (practically) suspended.[12] Sleep is different from a complete loss of conscious-ness, since you can be awakened by a noise or a shake.

Stages of Sleep

- Non-REM sleep consists of light sleep, stages 1 (N1) and 2 (N2); and deep sleep, stage 3 (N3).
- REM (rapid eye movement) sleep, aka "paradoxical sleep," is when the body is paralyzed but the brain is almost as active as when it's awake. This is when you dream.

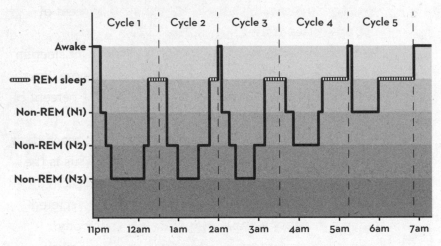

A "sleep cycle" is one rotation through the stages of sleep. Most people go through four to six sleep cycles a night, and each one lasts about 90–120 minutes. It's useful to note that wakefulness is part of the sleep cycle, too, and composes about 5 to 10 percent of sleep. These periods of wakefulness are called "partial arousals." That's when you change positions and, hopefully, go right back to sleep afterward.

It is possible that these partial arousals had an evolutionary purpose. Sleeping is dangerous when predators are around, so for our early ancestors, waking during the night may have been an important safety feature. Today, it's just part of a normal, healthy rhythm.

Each sleep stage has a distinct physiological and neurological function:[13]

1. Partial Arousal: About 5 to 10 percent of your nights are spent partially awake.
2. N1: This stage starts with drowsiness, heavy eyelids, slowed heart rate, eye movements, relaxing muscles, and perhaps occasional movements or jerking of the arms and legs. It ends with a light sleep. N1 is about 5 percent of your total sleep time.
3. N2: This stage starts with a deepening of the light sleep in N1. Heart rate and breathing slow down even more. Your muscles relax even further. N2 is about 45 to 50 percent of your total sleep time.
4. N3: This is deep sleep. Heart rate and breathing reach their slowest pace, and muscles are totally relaxed. This is the most difficult stage to wake up from. Your brain activity switches to delta waves. Growth hormones are secreted from the pituitary gland during N3. Also, your body restores physical energy by directing blood flow away from your brain and toward your muscles. N3 is about 15 to 20 percent of your total sleep time.
5. REM: During the rapid eye movement stage of sleep, your brain is very active—your breathing and heart rate become more irregular. This is when you dream. According to a recent study, everyone dreams, even though some people think they never do.[14] For some reason, every brain needs to dream. During REM sleep, the voluntary muscles in your body become temporarily paralyzed (so you don't act out your dreams). This stage is important for learning and

memory and accounts for about 20 to 25 percent of your total sleep time.

CONSEQUENCES OF POOR SLEEP

Current research suggests that, during sleep, a network of vessels called the glymphatic system washes away the neurotoxins and waste products that accumulate in the body when we are awake. During sleep, the space between the cells in the brain increases by 60 percent to allow for the clearance of these waste products, including something called beta-amyloid. The accumulation of amyloid plaques is one of the hallmarks of Alzheimer's disease. It certainly makes one wonder what happens to a brain that isn't adequately "washed" during sleep.[15]

Here are some other known examples of the problems that can arise when we don't get enough quality sleep:

- **Immune System Function**
  Studies have shown that sleep deprivation can decrease the immune response to the flu vaccine and increase the risk of catching the common cold by 300 percent.[16]

- **Learning and Memory**
  When you learn something new, you need to store it as a lasting memory. There is a new body of evidence that suggests that sleep plays an important role in this storage process.[17] This is true for facts and experiences, but also for physical skills like riding a bike or playing an instrument. We also know how much sleep can be helpful for finding creative solutions to problems. Think about how many times you've awoken with a solution to the problem you were considering when you fell asleep. The saying "Let's sleep on it" is popular for a reason!

- **Alertness and Attention**
  A lack of sleep affects a person's behavior, appetite, and ability to pay attention. This is most obvious in kids. The National Sleep Foundation found that 15 percent of children fall asleep at school because of a lack of sleep at home. That's three kids in a class of 20! Falling asleep in class is extreme, but I often wonder how many kids are just tired enough to struggle to pay attention. In a study of 263 children over a five-year period, the children with persistent sleep apnea were seven times more likely to have learning problems and three times more likely to have school grades of a C or lower.[18] Researchers at the University of Chicago revealed that obstructive sleep apnea may reduce a child's IQ by as much as 10 points.

- **Behavioral Issues**
  In a landmark study by the National Institutes of Health (NIH), over 11,000 children with sleep-disordered breathing (snoring, apnea, and mouth-breathing) were examined from infancy through age seven. By age four, these children were 20–60 percent more likely to exhibit behavioral difficulties. By age seven, they were 40–100 percent more likely. The worst sleep symptoms were associated with the worst behavioral outcomes, especially hyperactivity. These children were also more likely to experience anxiety and depression, have problems getting along with peers, and show more aggression.[19]

    As any parent knows, tiredness causes children to have explosive tempers, easily hurt feelings, and impatience. Sleepiness is especially unsafe in young children, who tend to become clumsier and more accident-prone and, at the same time, more restless and frenzied. It's a dangerous combination. This is why, when compared to their well-rested peers, children ages three to five who go more than eight or nine hours without sleep are 86 percent more likely to visit the emergency room.[20, 21]

WHAT IS SLEEP-DISORDERED BREATHING (SDB)?

Sleep-disordered breathing (SDB) is the spectrum of sleep-related issues, with habitual snoring at one end and obstructive sleep apnea at the other.

With normal breathing, there are no obstructions to the flow of air through the nasal passages. It is very quiet and easy. Each breath is full and clear. Normal breathing during sleep is deeper and slower, but it's still easy and clear. There is no snoring, gasping, coughing, sniffling, or other airway-related discomfort.

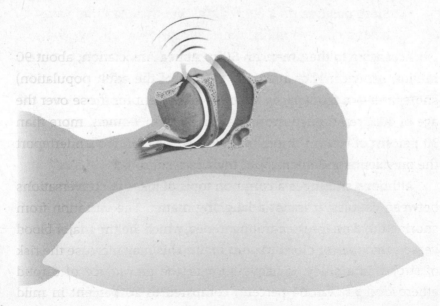

Obstructed breathing, especially in adults, usually involves some kind of snoring. Snoring is noisy sleep breathing caused by the vibration of soft tissues in the nose and throat. It happens when airways are too narrow or are obstructed in some way.

Snoring can be caused or worsened by many factors, such as:

- Macroglossia (large tongue)
- Low draping soft palate
- Enlarged tonsils/adenoids
- Nasal obstruction
- Small jaws
- Collapsible airway
- Aging
- Gaining weight
- Excessive fatty tissues
- Genetic predisposition
- Epigenetic factors
- Alcohol or medications
- Sleep position

According to the American Sleep Apnea Association, about 90 million Americans (or about 40 percent of the adult population) snore, and that figure rises to about 60 percent for those over the age of 40. Even though more men snore than women, more than 30 percent of women snore too. Women also tend to underreport the prevalence and intensity of their own snoring.[22]

Although snoring is a common topic of jokes in conversations between couples, it is not a laughing matter. The vibration from snoring can damage the carotid arteries, which are the major blood vessels that deliver blood to your brain. This may increase the risk of stroke. In a study of heavy snorers, the prevalence of carotid atherosclerosis was 64 percent, compared to 20 percent in mild snorers. Additionally, if snoring is not addressed and worsens, it can be a significant risk factor for developing sleep apnea, which is a very serious medical condition. Between 50 to 70 percent of middle-aged, habitual snorers already have this debilitating problem (more on apnea later).[23]

TIPS TO REDUCE OR ELIMINATE SNORING

- Lose weight.
- Get more physical activity during the day.
- Limit alcohol consumption before going to bed.
- Avoid taking sleeping medications.
- Sleep on your side.
- Take a nasal decongestant when you are congested.
- Do breathing and tongue exercises (see pages 85 and 86). Retraining your daytime breathing can be helpful for your nighttime breathing too.

If your snoring doesn't resolve with the tips above, please see a sleep physician or a sleep-trained dentist. It is critical for your long-term health.

Nobody should snore, but that's especially true for children, so please seek treatment if you notice your kids snoring. Again, airway and sleep disorders can have a profound impact on the growth and development of a child.

THE DESTRUCTIVE CYCLE OF SLEEP APNEA

The word "apnea" comes from the Greek *apnoia*. *A-* means "lack of," and *-pnoia* means "breath." A person with sleep apnea may fall asleep just fine. In fact, they are usually so exhausted, falling asleep is rarely a problem. But as soon as the muscles and tissues of the face, neck, and mouth start to relax, the trouble begins. Those tissues block the airway and the body panics. Each time the lungs try to expand and can't, the body releases adrenaline, in an attempt to jolt itself awake. For the sleeper, this results in an unpleasant awakening: coughing or gasping for air. Meanwhile, the body is filled with adrenaline—a stimulating hormone—making it harder to fall back asleep. Eventually, the person does fall asleep again, only to have the whole exhausting cycle repeat, sometimes many times every hour.

To make matters worse, patients suffering from sleep apnea subconsciously avoid going to sleep because sleep is extremely unpleasant for them. They often gravitate to the couch or lounger in front of the television, where they can sleep much more comfortably, since their airway stays open more easily in a seated position. Of course, seated TV sleep isn't the most restful. Once they get up to go to bed, the oxygen deprivation/adrenaline/jolting awake cycle continues.

People with sleep apnea tend to feel very tired all the time because they never manage to get through the deeper stages of sleep that are so critical for brain and body repair. The typical diagnostic criteria for sleep apnea is that the patient experiences no breathing for at least 10 seconds during sleep, five or more times per hour. For some people, it happens much more often than that, and their sleep is disrupted hundreds of times every single night.

In addition to being exhausting, sleep apnea can cause a drop in the level of oxygen in the blood. To give you a sense of the severity of oxygen deprivation, even with moderate sleep apnea, patients will often have their blood oxygen drop from around 98 to 100 percent to 80, 70, or even into the 50 percent range.[24] For context, if your blood oxygen drops to 92 percent in a hospital, they will put an oxygen mask on you. When conscious, it is almost impossible to get your blood oxygen to drop to 90 percent by holding your breath.[25] When oxygen levels dip this low, it can have harmful effects on the cardiovascular system and the brain.

People with sleep apnea may have memory and concentration difficulties and be less productive throughout the day. They may be more prone to work injuries and driving accidents. There are too many potential consequences of this type of chronic oxygen deprivation to mention, but they range from obesity to cardiovascular disease, and from sexual dysfunction to Alzheimer's.[26]

Three Types of Sleep Apnea

- **OSA (obstructive sleep apnea):** This is the most common type of sleep apnea. An estimated 25 million Americans between the ages of 30 and 70 suffer from it.[27] The prevalence of OSA has increased substantially over the last two decades, partially as a result of the obesity epidemic.
- **CSA (central sleep apnea):** This type is less frequently diagnosed and can occur without snoring. It is caused by the interruption of signals from the central nervous system to breathing muscles.
- **Mixed (complex sleep apnea):** This is a combination of obstructive and central sleep apnea.[28]

SLEEP AND CHILDREN

Let me tell you a story.

An eight-year-old boy is physically small for his age. He has a relatively small mouth, some crooked teeth, and an overbite. He also snores loudly at night and sleeps with his mouth open. His parents have a hard time waking him for school, which he hates attending. His teacher calls a meeting with his parents because, in addition to his poor performance and trouble paying attention, he has started picking fights with other students. The teacher suggests an ADHD diagnosis and medication.

The kid happens to have a dentist appointment coming up, and the parents decide to ask about the crooked teeth. The dentist tells them it's fine because all kids have crooked teeth these days, and they should just wait until all the permanent teeth come in and see what happens. Maybe he'll grow out of it, and if not, they'll just pull some teeth to make room for the rest. The family's insurance won't pay for braces until the child is 12 years old anyway.

The boy's parents make an appointment with his pediatrician and, after reviewing his symptoms, the doctor agrees with the teacher and prescribes Ritalin to manage his ADHD. He also recommends the child see an ENT to take out his tonsils to fix his snoring, since they look quite large.

The truth is that this imaginary boy is based on one of my patients. His parents finally brought him in to see me for a second opinion. Together, we decided to use a combination of an expander on the upper jaw to create more space for his teeth and airway, and a functional device to help grow his lower jaw forward. This second device fixed his overbite and gave his tongue more room. Then he completed a series of tongue, breathing, and swallowing exercises to retrain his tongue posture and to learn how to breathe through his nose. This timely orthodontic intervention resolved *all* of his problems without any medication or surgery. He didn't have ADHD; he just needed better sleep and more oxygen in his brain.

Because we intervened to fix this little boy's breathing and bite, his life and future was set on a much healthier and more prosperous path. All of these changes helped him to sleep better, which, in turn, improved his behavior, attention, and school performance. And because he had deeper, more restorative sleep (during which growth hormones are released), he grew—a lot! Today, he is of above-average height. He is strong, happy, and enjoying his life. And he didn't spend his childhood years taking medication. Who knows how that might have affected his long-term development?

For young children, the connections between breathing, sleeping, behavior, growth, and development are clear, if you know to look for them. How well a child develops and learns depends so heavily on having a consistent, healthy amount of oxygen in the brain. Early interventions can mean the difference between a lifetime of struggle and a lifetime of health and prosperity.

Many cases of airway obstruction in children go undiagnosed (or misdiagnosed) by health care providers. This is largely due, again, to our fragmented medical system. Airway and sleep-related disorders may require collaboration between pediatric dentists, adult dentists, orthodontists, pediatricians, allergists, ENTs, surgeons, sleep specialists, and myofunctional therapists. To complicate matters further, only a small fraction of dentists and physicians are airway-focused and/or sleep-trained.

But really, parents are the key. Parents advocate for their children like nobody else. A well-educated parent will know what to look for and has a particular opportunity to observe and detect problems early. This is so important because children with sleep-disordered breathing may have a higher risk for so many frightening issues: SIDS, ADHD, systemic diseases, obesity, and behavioral challenges.

Here's another example:

Frances, the one-year-old daughter of a friend of mine, had always breathed through her mouth. She was a healthy, happy baby, but she snored and sometimes seemed to struggle to nurse. When Frances's teeth started to come in, they were crooked and crowded. But her pediatrician was not concerned. He shrugged off my friend's questions and suggested waiting. He said Frances didn't need to go to the dentist until she turned three, and they should just see if the problem resolved on its own. My friend was not satisfied with this approach—in large part because she knew mouth-breathing was not normal and that it could contribute to dental crowding. Night after night, she listened to Frances's snoring and worried about what that lack of oxygen was doing to the child's developing brain. My friend, who lives in Minnesota, came to me for advice and I suggested that she take her daughter to a pediatric dentist, something kids should do before age one anyway—my friend's pediatrician was woefully misinformed about ADA recommendations. About five seconds into the pediatric dental appointment, Frances started to cry, and the doctor said, "Oh! She can't breathe through her nose. I can hear it in her voice."

Parents always know their babies better than anyone, but it can be so hard to insist on the kind of care you feel your child needs when you, yourself, are not a medical professional. Having your suspicions confirmed can be difficult—especially when those suspicions involve your baby's ability to breathe! But it can also be an enormous relief. My friend was so glad to be taken seriously, and to be able to investigate the cause of her daughter's breathing troubles. The dentist referred my friend to an ENT (ear, nose, and throat doctor) who discovered a 95 percent nasal blockage from enlarged adenoids. They were almost completely preventing little Frances from breathing through her nose. Any inflammation—from dust or a minor cold or dry air—and Frances's nose was completely blocked.

Frances's adenoids have since been removed, and she is now breathing much easier. She no longer snores and sleeps much more soundly because of it.

SIGNS OF SLEEP-DISORDERED BREATHING IN CHILDREN

Although snoring and mouth-breathing are the most common indicators of airway obstruction or SDB, there are many other signs you should look for in children:

- **Behavioral**
  - Attention problems
  - Tiredness and difficulty arousing
  - Crankiness
  - Bullying or aggression
  - Developmental problems
  - Poor school performance
  - Night terrors or nightmares
  - Nail-biting
- **Physical**
  - Mouth-breathing (during daytime and sleep)
  - Snoring or noisy sleep
  - Crooked teeth (dental crowding)
  - Narrow, high palatal vault
  - Congenitally missing teeth (underdeveloped mouth)
  - Bite issues such as a crossbite, open bite, underbite, or overbite
  - Grinding teeth at night (bruxism)
  - Cracked/dry lips
  - Bad breath
  - Dark circles around the eyes
  - Drooling at night (wet pillow)
  - A nasal voice

- Tongue issues including tongue-tie or poor tone
- Enlarged tonsils and adenoids
- Chronic ear infections
- Chronic sinus infections
- Allergies
- Using facial, neck, or chest muscles for breathing during sleep
- Interruption of breathing during sleep or gasping for air
- Bed wetting
- Lots of movement during sleep; messy sheets in the morning
- Chronic runny nose
- Headaches
- Failure to thrive
- Poor physical growth
- Obesity (which in turn makes sleep apnea worse)

SLEEP AND ADHD

Diagnoses of ADHD (attention-deficit/hyperactivity disorder) have significantly increased in the past several decades.

Since sleep deprivation and sleep disorders have very similar symptoms to ADHD, patients are often misdiagnosed. This is quite alarming because so many of these patients are medicated with stimulant drugs that can make sleeping even more difficult. It's true that patients with ADHD have a significantly higher rate of sleep problems: one out of four children had some type of sleep-disordered breathing.[29] As grown-ups, sleep deprivation typically manifests itself as exhaustion or crankiness. We slow down. But for kids, "tired" means "wired." If you have young children or have been around young children up past their bedtimes, you know what I'm talking about.

For example, my family and I recently got back from an overseas trip. My wife and I were jet-lagged and walking around the house like zombies, but our son was running in circles, jumping on the furniture, and screaming at the top of his lungs. My wife and I didn't even have the energy to ask him to quiet down. And this was just from one night of interrupted sleep.

If your child has been diagnosed with ADHD, I would encourage you to consult with a sleep-trained dentist or physician to rule out sleep-disordered breathing. It could be as simple as expanding the upper jaw or positioning the mandible forward. Imagine if a simple orthodontic device could save your child from a lifetime of medications!

SLEEP AND BULLYING

Bullying is one of the biggest worries that keep parents up at night. StopBullying.gov reports that 20 percent of U.S. students say they have been bullied at school, and more than 15 percent have been cyberbullied. These children have a higher risk of developing social and psychiatric symptoms later in life, too, including delinquency, substance abuse, violence, antisocial behavior, and criminal activity.

Studies have shown that one possible biological contributor to aggressive behaviors may be sleep-disordered breathing (SDB).[30] Evidence is now growing in support of the idea that SDB may cause or contribute to disruptive behavior disorders. In an inconvenient twist of fate, when it comes to bullying, the mouth may be responsible for children who bully, and for the kids they pick on. The number one physical feature children get bullied about is their teeth.[31]

SLEEP AND OBESITY

Since the 1970s, the percentage of children and adolescents affected by obesity has more than tripled.[32]

Sleep and obesity are closely linked (more on this later). In a study of 915 babies, those infants who slept less than 12 hours a day had twice the likelihood of being overweight when they were three years old.[33] In another study of more than 8,000 children, those who slept less than 10.5 hours a night at age three had a 45 percent higher likelihood of becoming obese by age seven.[34] Like bad teeth, obesity is a common reason why kids are teased and bullied. And it's something that can cause many health-related problems as kids grow.

SLEEP AND SIDS

Sleep apnea has been suggested as a potential contributor to SIDS (sudden infant death syndrome).[35] In a recent study, researchers Caroline Rambaud and Christian Guilleminault investigated the anatomical and sleep history risk factors that are connected with abrupt sleep-associated death in children. Medical histories revealed the presence of chronic indicators of abnormal sleep in all of the cases they studied. All cases demonstrated variable enlargement of upper airway soft tissues, and there were features consistent with a narrow, small nasomaxillary complex. The children were concluded to have died of hypoxia (oxygen deficiency) during sleep.[36]

HOW TO IMPROVE AIRWAY OBSTRUCTION AND SDB IN CHILDREN

When it comes to airways and breathing, prevention, early detection, and treatment should be top of mind for every parent. As discussed earlier, because an infant's life is dependent upon healthy breathing and a child's brain is dependent on oxygen during its explosive growth, airway and sleep difficulties can have dramatic, sometimes irreversible, consequences for children. Like everything else, it's always best to try to prevent snoring and sleep apnea, rather than waiting or treating it in adulthood. When signs of these problems surface in children, prompt treatment is crucial.

Remember, a healthy mouth, healthy body, well-formed face, and open airways all go together. A healthy pregnancy, breastfeeding, a nutritious diet, and healthy habits such as nose-breathing can all help grow a child's mouth and airways correctly and prevent problems later. All of this is connected.

TREATMENT OF PEDIATRIC SDB

The first step to treating sleep-disordered breathing is to address any allergies, asthma, or nasal congestion that may be preventing proper oxygen intake. Once those issues have been resolved, you can move on to tackling the breathing. Treatment options vary depending on root causes, but in general, the earlier you detect and treat problems, the easier, cheaper, and more effective treatments tend to be. Younger children are more susceptible to potential problems, but they are also more responsive to treatments.

Here are a few possible options:

- Orthodontic treatment. In some cases, maxillary expansion may be enough. In others, you may need to pursue a combination of orthodontic treatments.
- Adenotonsillectomy. If they're significantly enlarged or chronically infected, your health care professional may consider removal of the tonsils and/or adenoids.
- Myofunctional therapy. This can help correct the improper function of the tongue and facial muscles through exercises and instructions.
- A positive airway pressure or CPAP device. These can help in severe cases.

WHICH DOCTOR CAN HELP?

Children's faces and mouths are still growing, which makes treating them fundamentally different from treating adults. Growth during treatment can be helpful, but it can also present some complications. Growth is helpful because we can modify or mold it as it happens. We can also retrain muscles and correct harmful habits much more easily when kids are young. But growth of the jaws, lymphatic tissues (tonsils and adenoids), the head, and the body are all disproportionate to each other. It's important that your doctors consider the developmental stage of the child and the future growth of tissues.

Dentists (can and should) play a significant role in the early detection and treatment of airway obstruction and SDB. In 2017, the American Dental Association (ADA) adopted a policy on dentistry's role in the treatment of SDB. It emphasized that all dentists should screen for sleep-breathing disorders and that "dentists are the only health care provider with the knowledge and expertise to provide oral appliance therapy (OAT)."[37, 38] Dentists refer to physicians for diagnosis and work with a multidisciplinary treatment team of orthodontists, myofunctional therapists, sleep specialists, and ENT doctors, when necessary.

Below are some examples of what each practitioner can do to help, though their roles may overlap or change depending on the individual case.

- **Pediatric Dentists**
  Airway-focused pediatric dentists are uniquely positioned to identify patients with an increased risk for SDB. They are often the first to notice symptoms. They can also take the lead in the coordination of care between the different specialists involved.

They can order sleep tests and work with sleep physicians when they diagnose a sleep disorder. And they can help with oral developmental problems such as releasing a tongue-tie. Studies show that an untreated short lingual frenulum (tongue-tie) is associated with obstructive sleep apnea (OSA) at a later age.[39, 40] If your child has a tongue-tie, your pediatric dentist should be screening for OSA. Pediatric dentists can also help the child build better oral habits.

- **Sleep Physicians**

  Sleep physicians are specialists in sleep and sleep disorders such as sleep apnea, insomnia, restless leg syndrome, periodic leg movement disorder, and narcolepsy. They can conduct sleep studies, observing patients during sleep to witness symptoms present firsthand.

- **Orthodontists**

  I always tell my patients: if I were the only orthodontist on a deserted island and I could only take one orthodontic device with me, I would take an expander (rapid palatal expander or rapid maxillary expander). Yes, these are the things orthodontists think about. An expander is an orthodontic device that typically attaches to the upper teeth and gently expands the upper jaw over a few months. When an orthodontist uses it correctly, an orthodontic expander can create more space for the erupting permanent teeth and can reduce or remove the need for extractions, which helps keep the jaws wide and airways open. It can also help create broader, nicer smiles, widen the jaws, correct crossbites, and even help with correction of other bite issues such as underbites and overbites. It can increase the size of the airway by widening the palate, which, in turn, widens the nasal cavity and helps with breathing and sleeping.

Orthodontists can also guide the growth of the mandible forward using functional appliances that add more space to the lower airways. Straightening teeth may also be helpful, as straight teeth help keep the tongue in the right position. Having front teeth in the right position allows the lips to come together gently, and that can help with nose-breathing too.

Orthodontists can also help correct habits such as thumb-sucking, tongue-thrusting, and mouth-breathing, though some of these habits are in the domain of the myofunctional therapist.

- **Myofunctional Therapists**
Myofunctional therapy is a noninvasive adjunct treatment for sleep breathing disorders. The therapists help retrain muscles through breathing and muscle exercises. Oropharyngeal exercises have been shown to significantly reduce the severity and symptoms of OSA.[41] Adding myofunctional therapy improves outcomes for children whose tonsils and adenoids are removed. The absence of myofunctional therapy treatment is associated with a recurrence of SDB.[42, 43]

- **ENT**
An ENT is an ear, nose, and throat doctor. This is the person who can perform an adenotonsillectomy (removal of the tonsils and adenoids). The tonsils are the two lymphatic tissues in the back of the throat, filled with immune cells. Adenoids are a little higher in the throat behind the nose and the soft palate. Tonsils and adenoids are the body's first line of defense. They sample and filter out bacteria and viruses that enter the body through the nose or mouth. Sometimes, they can become infected and enlarged, obstruct the airway, force a person to breathe through their mouth, and interfere with sleeping.

IS REMOVING TONSILS AND ADENOIDS THE RIGHT SOLUTION FOR SDB?

When a child breathes through their mouth, the increased turbulence of the breath on the throat can cause inflammation and enlargement of the tonsils and adenoids. This creates a vicious cycle: mouth-breathing fails to humidify and filter the air, which leads to enlargement of tonsils and adenoids and affects breathing and sleep quality. A lack of quality sleep increases the risk of catching a cold or getting a throat infection, which further enlarges the tonsils and adenoids.[44, 45]

Recurrent infections and the enlargement of these tissues can cause airway obstruction and sleep problems. However, enlargement of the lymphoid tissues may be a consequence of airway obstruction and mouth-breathing, as opposed to a cause. In these cases, teaching mouth-breathing children to breathe through their nose can potentially reduce the inflammation, especially if the damage has not been chronically occurring.

Even though removing tonsils and adenoids is common practice, the root cause of OSA may be related to abnormal oral-facial growth, particularly during early childhood. This won't be addressed by adenotonsillectomy. The debate over removing tonsils and adenoids—similar to the debate over pulling permanent teeth for braces—has gone on for decades, during which the pendulum has swung back and forth. A doctor should conduct a thorough evaluation of the dynamics leading to the development of SDB and the factors affecting craniofacial growth before recommending surgery.[46]

Perhaps the question shouldn't be, "Should we take them out?" but rather, "When is it necessary to take them out?" Remember that tonsils grow disproportionately to the rest of the body. If you examine the tonsils on a 10-year-old, they will look abnormally large. A doctor should carefully consider age and stage of growth.

The severity of airway obstruction is also an important deciding factor: in many cases, since the damage to the developing body and brain of a child can become significant, waiting for the airway to get bigger after puberty may not be a good option.

Another consideration: Are there other ways to enlarge the airway? For example, before the maxillary sutures fuse (they start fusing around age eight), an orthodontist can expand the upper jaw and perhaps bring the lower jaw forward, to create more breathing room and reduce the need to surgically remove the tonsils and/or adenoids.

Having said that, sometimes it is necessary to do an adenotonsillectomy. In many cases, an interdisciplinary approach is required for the long-term treatment success of airway obstruction and SDB: an ENT can remove the tonsils and adenoids, an orthodontist can expand and align the jaws, and a pediatric dentist and/or myofunctional therapist can retrain the muscles to help correct habits and reestablish full-time nose-breathing.

Interestingly, Christian Guilleminault, one of the preeminent experts in this field, has concluded that "if nasal breathing is not restored, despite short-term improvements after T&A [tonsil and adenoid removal], continued use of the oral breathing route [mouth-breathing] will be associated with abnormal impacts on airway growth and possibly blunted neuromuscular responsiveness of airway tissues, both of which may predispose to the eventual return of upper airway collapse in later childhood, or in the full-blown syndrome of OSA in adulthood."[47]

All this is to say, even removing the tonsils may not work long term if nose-breathing isn't sustained. Even if tonsils and adenoids are removed, this is a great argument for myofunctional therapy, orthodontics, and other support. The end goal of all of this is to make sure nose-breathing becomes the norm.

SLEEP AND ADULTS

About half of the adult population suffers from some form of SDB, which means almost everyone is affected, because even if you don't snore, your partner might. It has also been shown that a snorer's bed partner loses an average of 62 minutes of sleep a night, to say nothing of how this might disrupt stages of sleep.[48] Bed partners can even have a higher risk for hearing loss in the ear closest to the person snoring.

According to the American Sleep Apnea Association, an overwhelming majority of apnea sufferers remain undiagnosed and untreated.

Incidence vs. Treatment of Obstructive Sleep Apnea

Diagnosed (10% of sufferers)

Undiagnosed (90% of sufferers)

OSA

Diagnosed and treated (25% of those diagnosed)

Total population

Population with OSA

Severe obstructive sleep apnea increases the risk of death from any cause by more than three times and can accelerate aging and reduce life expectancy by up to 12 to 15 years.[49]

Sleep is a beautiful thing for people who sleep easily and deeply. Sleep can also be a nightmare—literally—for those who have difficulty.

SIGNS AND SYMPTOMS OF SDB

Daytime

- Drowsiness
- Body aches
- Low energy
- Irritability
- Mood changes
- Anxiety
- Depression
- High blood pressure
- Low sex drive
- Erectile dysfunction
- Poor attention
- Learning difficulties
- Craving sugar, caffeine, or junk food
- Falling asleep sitting up, in front of the TV, when reading, or while driving
- Bad breath and/or dry mouth
- Bite issues: crossbite, overbite, underbite, open bite
- Small mouth: dental crowding, scalloped tongue
- A large tongue (macroglossia)
- A narrow or high palate

Nighttime

- Snoring
- Frequent tossing and turning
- Gasping for air in the middle of the night
- Frequent nightmares
- Tongue thrusting and a scalloped tongue
- Frequent awakenings at night
- Frequent urination
- Acid reflux or GERD (gastroesophageal reflux disease): When you stop breathing, your body increases its efforts to get your oxygen intake back on track using abdominal contractions to take in air. These contractions squeeze the stomach acid up the esophagus. This can also cause erosion of the tooth enamel over time.
- Clenching or grinding of teeth (bruxism)

Daytime (Cont.)

- Headaches: OSA can lead to three types of headaches:
  - Migraine—linked to disruption of REM and delta sleep patterns
  - Dull morning headaches—linked to oxygen deprivation
  - Cluster headaches—linked to oxygen deprivation[50]

Nighttime (Cont.)

- Wear on bite surfaces due to grinding
- Apnea causes the body to activate the jaw and tongue muscles to regain control of the airway and keep it open. It's important to note that patients who grind their teeth need to be evaluated for OSA before they get fitted for a splint or night guard because using one of these devices can actually worsen the airway obstruction during sleep.

Some of the symptoms are more obvious as they relate to SDB. Sleepiness during the day, for example, is usually a sign of poor sleep. Other signs, however, may not be so obvious. For example, frequent night awakenings to urinate may just make you think you're drinking too much before bed. Although that can definitely be *one* reason, your body may also be giving you a hint.

Generally, when you sleep well, your body relaxes and allows you to sleep through the night without the urge to urinate. But if you have obstructive sleep apnea, your body will wake you up as you struggle for air. You may not remember these episodes. It's just once you're awake that you feel the urge to urinate. SDB can also affect the secretion of urinary hormones: an increase in atrial natriuretic peptide (ANP) and a decrease in antidiuretic hormone (ADH) can prevent the normal decrease in nighttime urine output.

THE CONSEQUENCES OF POOR SLEEP IN ADULTS

Adults with sleep deprivation and sleep-disordered breathing can have an increased risk of work injuries, automobile accidents, poor job performance, faster aging, increased blood pressure, coronary disease, heart attack, stroke, high cholesterol, glaucoma, diabetes, weight gain, obesity, decreased libido, erectile dysfunction, insomnia, anxiety, depression, cancer, dementia, Alzheimer's, and increased mortality.

Yikes!

I don't say all of this to scare you . . . or maybe I do. Sleep-disordered breathing is a serious problem that, in my opinion, our culture and the medical and dental professions do not take seriously enough.

Of course, many of the consequences of sleep deprivation have multifactorial causes, and the correct diagnosis of the root cause is challenging. This is when coordination of care really counts.

Obesity

Globally, the rates of obesity have increased dramatically in both children and adults. Today, 17 percent of children and up to 40 percent of adults are considered overweight or obese. For Americans, the numbers are even more shocking: more than 70 percent of U.S. adults are overweight or obese.[51]

Insufficient sleep has harmful effects on the hormones involved in appetite control, leptin and ghrelin (the "hunger hormones").[52] Leptin decreases your appetite and ghrelin increases it. During sleep, leptin levels rise, telling the body it is not hungry.

Sleep deprivation disrupts the secretion of leptin, stimulating the appetite. This can help explain why people tend to want to eat when they can't fall asleep, or why they crave a juicy burger after a late night of partying.

Here we have yet another complex reality. All of these factors tend to affect each other. For example: too much stress (and stress hormones like cortisol) can contribute to obesity; obesity can make SDB worse and lead to insomnia; and insomnia can affect leptin and ghrelin, causing inappropriate hunger.

CHEW ON THIS

In addition to healthy eating, physical activity, and a balanced micro-biome, quality sleep must also be a part of everyone's long-term weight-management program.

Increased Blood Pressure

OSA increases the risk of hypertension by 45 percent. There are two reasons:

1. When the body's oxygen is depleted, there is a sympathetic (emergency) response to jolt the body awake. This increases heart rate and blood pressure. When these events happen hundreds of times every night, the body remains in a constant state of emergency and blood pressure stays elevated.
2. Insufficient oxygen intake and nitric oxide can lead to narrowing of the blood vessels and a breakdown of the endothelial lining of the arteries. This can increase the risk of high blood pressure, heart attack, and stroke.[53]

Heart Disease

Heart disease is the leading cause of death in the United States. Obstructive sleep apnea is associated with obesity and high blood pressure, which are both major risk factors for heart disease. The American Heart Association released a statement recently indicating that "sleep disorders such as SDB and insomnia are associated with adverse cardiometabolic risk profiles and outcomes."[54]

In a massive study published in the *Journal of the American College of Cardiology* tracking almost 11,000 people over a five-year period, researchers found that people with obstructive sleep apnea had a significantly higher risk of sudden cardiac death, and the risk increased with the severity of the apnea.[55, 56]

Stroke

Conclusive research in the *New England Journal of Medicine* in 2005 reported that sleep apnea "is significantly associated with the risk of stroke or death from any cause, and this association is independent of other risk factors, including hypertension. Increased severity of the syndrome is associated with an incremental increase in the risk of this composite outcome."[57] In another study of OSA patients published in *Sleep*, 71.9 percent had cardioembolic strokes, compared with 33.3 percent in non-OSA patients.[58]

Diabetes

Sleep deprivation can lead to an increase in resistance to insulin, affecting the regulation of glucose and lipids.[59] Meanwhile, a tired body secretes more stress hormones, such as cortisol, which may help you stay awake but also interfere with the action of insulin, further disrupting sugar metabolism.[60]

Microbiome Changes

Intermittent hypoxia (low oxygen) and hypercapnia (high CO_2) in obstructive sleep apnea alters the gut microbiome, which could also contribute to adverse cardiovascular and metabolic outcomes.[61]

Cancer

Studies show that sleep deprivation can be a contributing risk factor for a number of types of cancers, including prostate, colorectal, and breast cancers.

Growth Hormone

Growth hormone is released by the pituitary gland. In children it promotes growth and in adults it helps to maintain normal body structure and metabolism. It also repairs and maintains muscles, including the heart. Sleep apnea sufferers don't get enough deep sleep when growth hormone is released. Clinical features of growth hormone deficiency can include changes in memory and attention, anxiety, depression, fatigue, decreased muscle mass, decreased bone density, and impaired cardiac function.[62]

Immune System

Sleep has a strong regulatory influence on immune functions, and sleep deprivation can suppress the immune system, making it more likely to get sick and more difficult to recover from it.

Sexual Dysfunction

Nitric oxide plays an essential role in erectile function. A lack of nitric oxide (from SDB) leads to narrowing of arteries and a decrease in the blood flow to the sexual organs.[63]

Automobile Accidents

According to the National Sleep Foundation in 2020, 50 percent of American adults admit they drove while drowsy and 20 percent admit to falling asleep while driving.[64] A large study of 14,268 crashes between 2009 and 2013 showed that 21 percent of crashes in which a person was killed involved a drowsy driver.[65] Patients with sleep apnea who are untreated are two and a half times more likely to be involved in a car accident.

Dementia and Alzheimer's

In a joint 2017 study from Stanford University and Washington University School of Medicine, scientists found that even just one night of disrupted sleep is enough to raise the levels of amyloid beta linked to Alzheimer's disease. Additionally, they found several nights of sleep disruption raised another chemical linked to Alzheimer's, called tau.[66]

Scientists at UC Berkeley have found compelling evidence linking sleep deficiency with excessive deposits of the beta-amyloid protein believed to trigger Alzheimer's disease. For this reason, they reported that "sleep could be a novel therapeutic target for fighting back against memory impairment in older adults and even those with dementia."[67]

An even better approach would be to prevent sleep deprivation proactively, thus preventing associated plaque buildup in the first place. Researchers at New York University found that older patients with sleep apnea began experiencing cognitive decline about 10 years earlier than those without breathing problems. However, the good news from the study was that patients who were being treated for their sleep apnea delayed the onset of mental decline by the same 10 years, as well.[68]

Depression and Anxiety

Depression and anxiety have a bidirectional relationship with sleep. When you don't sleep well, it's more likely that you'll suffer from these disorders. And when you're suffering from these disorders, sleep often becomes more difficult. As the Anxiety and Depression Association of America puts it: "The more we look for sleep, the less we find it."[69]

UPPER AIRWAY RESISTANCE SYNDROME

UARS (upper airway resistance syndrome) is another sleep disorder in which resistance through the airway is significant enough to disrupt the quality of sleep. The increased effort to inhale during sleep can lead to many partial arousals (RERAs) throughout the night. However, this is different from OSA, in which breathing can completely stop. UARS can start as simple snoring, but if left untreated, it can lead to greater obstruction and eventually progress to OSA.

Healthy breathing   Snoring   UARS   Sleep apnea

What complicates the diagnosis of UARS is that patients may not be aware of their snoring (not all patients have audible snoring but are just "heavy breathers") and may not remember partial arousals. The clues may lie in the epidemiology and symptoms.

Both men and women can suffer from UARS, although the prevalence is slightly higher in women, especially young, thin women.[70] Many of the UARS symptoms are different from those of OSA. Patients with UARS may have difficulty falling asleep, be light sleepers, and have low blood pressure, so much so that patients may get dizzy or faint when they stand up quickly. They can suffer from severe fatigue and exhaustion, hypothyroidism, digestive issues like diarrhea, constipation, and bloating, anxiety and depression, and cold hands and feet (because the sympathetic nervous system sends more blood to the heart and away from the extremities—cold hands, warm heart!). Other symptoms may include: TMD (temporomandibular disorders), frequent nighttime urination, morning headaches, dry mouth upon awakening, jaw pain, and scalloped tongue.

In recent years, some studies have shown strong associations between UARS and functional somatic syndrome, which is a term used to refer to physical symptoms that are poorly explained and appear in otherwise healthy populations.[71] This includes disorders such as chronic fatigue syndrome, chronic insomnia, chronic pain, irritable bowel syndrome, fibromyalgia, and depression.

Treatment of UARS is very similar to treatment for OSA and should be considered the moment snoring has progressed and daytime symptoms are present.

Upper Airway Resistance Syndrome (UARS) vs. Sleep Apnea

	UARS	Sleep Apnea
Prevalence	Slightly higher in women	Higher in men
Sleep onset	Delayed, insomnia	Fast
Breathing	Resistance through the airway	Breathing completely stops
Snoring	Common	Almost always
Body type	Slim to normal build	Overweight
Daytime symptoms	Tiredness Fatigue	Sleepiness in adults Children can be "wired" or cranky
Orthostatic symptoms	Cold hands and feet Fainting Dizziness	Rare
Blood pressure	Low to normal	High

LIFESTYLE FACTORS CONTRIBUTING TO SDB

- Alcohol (especially drinking before bedtime).
- Sedatives or sleeping pills—similar to alcohol, these can relax muscle tone, leading to airway collapse.
- Smoking—smoke irritates and inflames the upper airway, causing it to narrow.
- Obesity—this can be a contributing factor to OSA, and also get worse because of OSA, creating a vicious cycle. Fat deposits on the neck and a larger tongue put pressure on the airway. The Mayo Clinic found that a neck size of 17" or more in circumference for men and 16" or more for women is a clinically significant risk factor for OSA.[72]

IS IT INSOMNIA OR OSA?

Insomnia is a recurring difficulty falling asleep and/or the inability to remain asleep for a reasonable amount of time. Chronic insomnia affects almost 20 percent of Americans.[73] It's a risk factor for substance abuse, depression, chronic pain, and many physical diseases. It may sound counterintuitive, but sleep deprivation can actually cause insomnia.

Shouldn't you sleep better if you're tired? Those of you who have had a baby probably know that when your baby misses a nap, they don't sleep better at night—they sleep worse. Constant sleep deprivation causes our bodies to produce stress hormones such as adrenaline and cortisol to keep us going. This can interfere with sleeping, creating yet another vicious cycle.

In a large study of 1,210 patients suffering from chronic insomnia published in *Mayo Clinic Proceedings*, it was found that 91 percent of the insomnia sufferers had moderate to severe sleep apnea.[74] What's truly frightening about this finding is that sleep medications, which relax muscles, can make apnea worse. It's likely that many of the people in this study had been misdiagnosed. How many of them took prescribed sleep medicine that made their underlying condition worse—even potentially lethal? What damage did that cause to their bodies? Imagine what happens when someone suffering from OSA takes sleep medication. Sedation prevents your body from waking you up. If you can't breathe and your body can't rouse you, you're in trouble.

If you're having difficulty sleeping or have chronic insomnia, please get yourself checked for OSA before you take sedating medication. It could save your life.

AIRWAY AND SLEEP ISSUES IN WOMEN

Even though SDBs are prevalent in both men and women, there are some unique characteristics that women should be aware of. Studies show that a significant majority of sleep-disordered breathing in women goes undiagnosed.[75] One reason for this could be that women's symptoms can differ from those experienced by men. Generally speaking, women are less likely to report snoring, but more likely to complain about fatigue, insomnia, morning headaches, nightmares, mood issues, anxiety, and depression.

Women are twice as likely as men to be treated for depression prior to getting diagnosed with OSA.[76] It has been hypothesized that in women, SDB may be an underlying cause of many anxiety disorders.[77] If you are a woman, and you feel fatigued, have insomnia, are overly stressed, or are suffering from anxiety or depression, you owe it to yourself and your loved ones to get evaluated for sleep-disordered breathing.

The latest estimate of women who suffer from moderate to severe OSA is up to 20 percent.[78] Typically, the prevalence of OSA is higher in men than women in their twenties and thirties, but that ratio shrinks to about two to one when people reach their fifties and sixties. This is because, as women age, their OSA risk increases. This phenomenon is related to hormone production, and to the way the body stores fat. Hormonal changes during pregnancy and menopause may also predispose women to SDB.[79] These figures may still be low—underestimating the prevalence of SDB in women who may not make the technical diagnostic criteria of OSA but still suffer from a partial upper airway obstruction such as UARS.

TREATMENT OF SDB IN ADULTS

Remember, when left untreated, OSA can take 12 to 15 years off your life. That's worse than smoking or diabetes.[80, 81] Sleep apnea has been coined the "not-so-silent killer," raising premature death risk by 46 percent.[82] It's never too late to take care of yourself. It's never too late to start eating healthy or start exercising. It's also never too late to improve your oral health so that you can breathe better and sleep better, no matter how old you are. You can't go back in time, but every day that passes is another missed opportunity to give yourself a better, longer life.

Remember that it is important that bed partners are part of the diagnosis and treatment planning of SDB, since they have both a vested interest in your health and unique insight into your symptoms. During your sleep consult, a qualified dentist or physician will review your medical history along with any oral or physical signs or symptoms. They will also evaluate the airway and may recommend a sleep study.

You can do a sleep study at home or in a sleep lab. Home studies map a typical night's sleep but offer fewer details and insight. Lab studies (polysomnography) are more detailed studies, but the night of sleep is less typical. You need to consult with a qualified dentist or physician to determine which will work better for you.

Once all the data are gathered, your sleep physician will recommend treatment options. It is important that you discuss your choices with your doctor and dentist. Compliance will be a key factor in determining which option will work best for you.

TREATMENT OPTIONS FOR ADULT SDB

- Lifestyle and behavioral changes such as weight management.
- Continuous positive airway pressure (CPAP). A CPAP pushes air down the airway to keep it open. CPAP is highly effective, but patients are frequently noncompliant with it, as it may be uncomfortable to wear at night.
- Oral appliance therapy. There are more than 100 oral appliances that have received FDA clearance. They keep the airway open by preventing the lower jaw from falling back during sleep. They can be very effective, and they typically have higher rates of compliance.
- Surgery. Removing the uvula and tissue from the soft palate is the most common but tongue, nose, and jaw surgeries are options as well. Soft-tissue surgeries tend to have lower success rates and are irreversible.

While you wait for that sleep consult, here are some other ways to improve your sleep at home:

- Establish a regular bedtime and wake-up time.
- Avoid caffeine and sugar.
- Use the bedroom only for bedroom activities.
- Avoid bedtime carbs.

- Avoid nighttime fluids to reduce restroom trips.
- If possible, do your activities during the day.
- Manage lighting (including reducing screen time before bed).
- Manage room temperature (cool, comfortable).
- Create a regular quiet, soft, and boring pre-sleep routine.

CONNECTING THE DOTS

Snoring, excessive daytime sleepiness, and hypertension (high blood pressure) are three of the most common symptoms of SDB. However, ask yourself the following questions:

- Are you often tired, fatigued, or sleepy during the daytime?
- Do you feel irritable or cranky?
- Do you have difficulty concentrating?
- Are you constantly craving sugar, caffeine, or junk food?
- Do you have, or are you being treated for, high blood pressure?
- Do you grind your teeth?
- Do you suffer from headaches?
- Do you hate the idea of going to sleep?
- Do you frequently wake up at night?
- Do you have difficulty breathing through your nose?
- Do you snore, or have you been told you snore?
- Has anyone observed you stop breathing during sleep?
- Is there a family history of SDB?
- Are you breathing through your mouth when you sleep?

Make sure you ask these questions of your bed partner too; they may respond differently than you think.

CHAPTER 4

THE POWER OF YOUR SMILE: PSYCHOLOGICAL AND EMOTIONAL HEALTH

IN THIS CHAPTER

- The critical role of the mouth in mental health.
- The emotional effects of poor oral health in children's lives.
- The mouth as a tool for personal and professional success.

How do you tell someone you like them? You could write them a letter or a song. You could buy them a gift. Or you could just smile.

There is a whole world in a smile: love, welcoming, joy, laughter, and friendship. Smiles lower barriers between people, open conversations, and welcome reciprocity. They show kindness. A genuine smile says, "You are welcome, and all is well." What a powerful message in a world that can be so divided.

Studies show that genuine smiling is associated with a better life. One study looked at college students' yearbook pictures and found a strong association between those smiling with positive emotional expressions in pictures and lower rates of divorce, as well as greater personal well-being up to 30 years later.[1] This may seem obvious: happier people smile more. But smiling also makes people happier. It's a chicken/egg situation.

Amazingly, it's true that the act of smiling sparks the happiness it represents. Put simply, by smiling, you tell your brain to be happy. Have you ever heard that you should smile when you feel sad? That it might help you to feel better? According to behavioral psychologist Sarah Stevenson, smiling is like a "feel-good party in your brain."[2] It triggers the release of dopamine, endorphins, and serotonin. In addition to elevating mood, endorphins are pain relievers. Smiling can reduce both pain and stress. And, since a smile is contagious, the smile you get in return triggers all those feel-good chemicals in other people too. Every time you smile at someone, you're supporting that person's mental health.

We smile to project feelings, and to inspire them in others. But smiling is so much more. Smiling can make you appear younger, thinner, and smarter. One smile can induce more brain stimulation than 2,000 bars of chocolate (without any of the calories or sugar).[3] People who smile more live longer.

Put it all together and you'll start to see that your smile is one of the most powerful things about you. It's the confidence and strength you project to the world, and to yourself. It profoundly affects how you feel.

In my practice, creating healthy, beautiful smiles is a daily occurrence. But when parents are deciding whether or not to fix their child's smile, some ask: "But, Doctor, isn't this just for looks?"

As you've learned, when it comes to the mouth, nothing is *just* for looks. The structure of the mouth affects the structure of the face, which affects breathing, oral health, and overall health. How you breathe, whether or not you have gum disease, how well you can chew and speak, and whether or not you have a chronic illness can all be related to a well-formed smile.

LOW SELF-ESTEEM AND THE MOUTH

The first five years of a child's life are arguably the most important. The explosive growth of the brain during this time has a lasting impact on personality, learning capacity, self-esteem, and confidence. At birth, the brain is about 25 percent of its final size. By age one, it doubles in size. By age three, it is 80 percent grown. And by age five, it is about 90 percent of its final size.[4] Any airway obstruction and oxygen deprivation during this crucial period of development can have long-term consequences.

Experiences during this developmental period—both positive and negative—help shape the way the brain grows too. This period is when children are learning about themselves, and the impact of peers on self-esteem is tremendous. This is when kids are developing a sense of self-worth, and when they are building their belief systems and values, figuring out what and who are good and bad.

This is another reason why the "wait and see" approach can be so detrimental. It may make physical treatment harder, and it also does nothing to undo the psychological damage that may have been done in the interim.

Kids suffer tremendously from the consequences of poor dental health, and the resulting insecurities can last a lifetime. Children with decayed or missing teeth hide their smiles and have difficulty speaking correctly. And their peers are quick to notice. These kids may begin to feel self-conscious and embarrassed. They may choose to stay silent rather than call attention to their teeth. This may hinder their development in subtle and not-so-subtle ways, from underdeveloped social skills to damaged self-esteem.

In my practice, I've seen many young children with low self-worth have a difficult time articulating their feelings. Sometimes they're labeled as shy or withdrawn because they keep to themselves. But the reality (as is often the case) is a lot more complex. I experienced that firsthand with a seven-year-old patient named Sofia.

One day, right before Sofia's first visit with me, her aunt asked to talk to me in private. She told me that Sofia and her family had been in a car accident a few months back. Sadly, Sofia's parents passed away, and now Sofia's aunt was her full-time guardian. Sofia had always been quiet, but the accident had left a few scars on her face, making her even more withdrawn. She told me Sofia felt lonely at school and had a difficult time making friends. And then she told me that Sofia had not been to a dentist . . . ever.

When I walked into the treatment room to examine Sofia, I caught a glimpse of a couple of beautiful drawings she had done before she quickly tucked them away. I introduced myself and extended my arm for a handshake, but Sofia just looked down to the floor and didn't say a thing. Sensing the awkwardness, her aunt quickly said, "Sofia is a little shy." To which I replied, "That's okay. From my experience, shy people are often incredibly smart, and from what I could see in her drawings, Sofia is also incredibly talented." Then I went on explaining how I was an artist, too, except my art was making beautiful smiles. That made her look at me finally and we started showing each other our artwork—she showed me her drawings and I showed her a few before-and-after photos of patients.

Sofia's art was truly impressive—way beyond what a seven-year-old (or even a 17-year-old) would typically draw. But there was a sense of sadness expressed in all of her pictures. Once she had become somewhat comfortable, I approached her to examine her mouth. The scars around her mouth made her skin so tight that I could barely get a dental mirror in there.

The years of neglect were immediately obvious. Sofia had rampant tooth decay, missing and crooked teeth, and an overbite. I was overwhelmed with sadness and excitement at the same time—sadness for what I saw, and excitement for what I knew I could do for her. During the next 18 months, I watched the transformation in her smile and in her personality happen in unison. And I could see that change in her art. Dark, sad drawings gave way to more colorful, happy ones. When she finally moved away a few years later, she had grown into a confident girl with a future full of possibilities ahead of her. I kept the last drawing she gave me in my office, and I cherish it to this day.

It may come as no surprise that the number one reason children are singled out for bullying is crooked, discolored, or rotting teeth.[5] Kids who have underbites may be teased for looking mean. Kids with overbites or those who breathe through their mouths may be teased for looking stupid: they're "mouth-breathers," after all. Sadly, the negative psychological and school-related consequences can haunt these children throughout their entire lives.

But dental-related problems with self-esteem and self-confidence often start more insidiously and quietly—with pain and discomfort, a lack of sleep, and other unrecognized symptoms. And often, the cascade of negative impacts can result in behavior problems, including bullying! So, unhealthy teeth and mouths can be to blame on both sides of the bullying equation. Eventually, the seeds of poor self-image can result in full-blown mental health crises: depression, anxiety, and even suicidality.

Over the years I have also seen many young children suffering from toothaches who have difficulty expressing their discomfort. Before being diagnosed, they simply endured their agony, not
knowing that it could be corrected. Kids who have always felt pain
may not know what it's like to live pain-free. And the poor parents,
as close as they are to their kids, had no idea that their children
were suffering.

Whether or not anybody knows about it, chronic pain can
interfere with a child's ability to sleep, affecting behavior, appetite,
and concentration. Tired kids are prone to explosive tempers, easily hurt feelings, and accidents.

Children with toothaches have difficulty eating, too, and are
likely to avoid hard-to-chew fruits and vegetables—some of the
most important foods for proper development.

Without a doubt, lack of sleep, poor nutrition, and pain cause
children to perform badly in school. In a study of school-age children, those who needed dental care were three times more likely
to miss school, and those who reported having toothaches were
approximately four times more likely to have a low grade-point
average (below the median GPA of 2.8) when compared with children who were free of dental pain.[6]

Poor academic performance in the early years can have a dom-
ino effect on education and is one of the strongest predictors for
continuously underachieving in school.[7] U.S. children annually
miss more than 51 million hours of school due to dental problems.
And, of course, children with toothaches or sleep deprivation find
it difficult to sit still and listen. I would too. Wouldn't you?

And so begins many a child's life in America. Think about what
often happens to kids with attention or behavioral problems: They
get medicated, treated badly, or teased. They start to believe that
their inability to concentrate is because they're deficient or stupid.
If they could just *try harder*, they are constantly told, they could

do better, and yet they just can't seem to get it together. They feel sad, lonely, and worthless. As they grow, the problems compound. They withdraw or act out, become bullies, turn to drugs, or get into trouble with the law. It's tragic because it didn't have to be that way. If you are a parent, you can break this negative cycle by taking your child to the dentist regularly, by making sure your child brushes and flosses, and by helping them build positive associations with their oral health.

Not only is it critical to care for kids' mouths, but, as you've now heard multiple times, childhood is the best (and cheapest) time to address problems. Kids' mouths are flexible and malleable. Often, if treated early, mouths and teeth will begin to grow correctly with minimal intervention. As the saying goes, "An ounce of prevention is worth a pound of cure." In light of all that you've just learned about the incredible power of a healthy smile, it should be getting clearer just how far a little prevention can go.

THE MENTAL HEALTH EQUATION

Many factors contribute to our mental health, and we're constantly learning more about the profound role of oral health in the equation.

As with many mental-health-related issues, the teen years are often when we really start to see things like clinical depression and anxiety emerge, but as we've discussed, the seeds are usually planted much earlier.

Today's children face a plethora of challenges that force them to grow up too quickly. Before they are ready, they may come face-to-face with divorce, death, incarceration, domestic violence, drug and alcohol abuse, racism, financial difficulties, and a host of other scary things. Many suffer from the pressure to overachieve, including the ever increasing need to perform well in school and participate in a never-ending torrent of extracurricular activities.

As a result, today's overwhelmed teens report being even more stressed than adults, with such negative results as impaired memories, weakened cognitive control, reduced learning ability, and poor emotional control. Subpar dental care, with all its negative impacts—loss of self-esteem and confidence, bullying, pain from toothaches, and difficulty breathing, speaking, and sleeping—only adds to a young person's ratcheting stress levels.

Psychological health can be hard to track. But we do know that low self-esteem and self-confidence correlate to negative outcomes: increased rates of anxiety and depression, poor behavior in school, acting out, subpar academic performance, and increases in substance abuse, eating disorders, teen pregnancy, and suicide rates. Addressing issues as early as possible is crucial.

IT'S ALL IN THE TIMING (AND THE DETAILS)

Orthodontic treatments in childhood—for tongue-ties, narrow palates, airway obstruction, bite problems, and correcting habits such as thumb-sucking—can help guide the healthy growth of the mouth and the face and circumvent many of the damaging psychological consequences of bad teeth. Sometimes braces can be entirely prevented (or at least, the severity of problems and treatment time can be reduced) with early interventions. Not only is that good news for the child, it's also good news for cash-strapped parents.

Straightening teeth should be the very last thing an orthodontist does, though often it's the first—and only—thing. Often, children simply arrive at the orthodontist too late, long after their bones have fused and skeletal growth is complete, when straightening crowded teeth is the only option. Sometimes parents and orthodontists are concerned with the aesthetics of straight teeth but don't appreciate how the underlying skeletal structure of the mouth can impact the end result.

Details really matter. Think about it: There are more than 7.8 billion people on this planet. We all have eyes, noses, lips, cheeks, chins, teeth, and jawlines in *almost* the same position; yet we all look so different. It's all about those minute variations.

A skilled orthodontist must start early with a growing child to guide the healthy growth and development of the face: the airway, bite, and oral habits. He or she needs to understand how to balance a face, and also imagine how that face will look post-orthodontics.

Once all of that has been taken care of, straightening the teeth puts the final touches on a healthy, beautiful smile. I always tell my patients, other than the parents, the person who will really affect what a child looks like for the rest of their life is the orthodontist. Of course, you cannot select your parents, but you can definitely select your orthodontist.

THE EYES HAVE IT

If your smile is the first thing someone notices about you, your eyes are the second thing. Studies show that people with nice smiles also tend to have nice eyes, a correlation that is both physical and psychological. As mentioned earlier, the bones and soft tissues that affect the look of the mouth (especially the maxilla—the middle third of the face) also affect the cheekbones and positioning of the eye sockets, and as a result, have a big impact on how the eyes look.

Psychologically, when you smile with confidence, your eyes reflect your inner feelings. If you're self-conscious about your mouth, you may be less likely to make eye contact—something that expresses confidence and helps build connections. Once again, it's all related.

THE ELEPHANT IN THE ROOM

I'm now going to broach a subject that might make you uncomfortable. But I would be remiss (and willfully oblivious) if I didn't discuss the topic of aesthetics.

We don't like to admit it, but in our society, the way we look is important. Whether we like it or not, beauty and success (both personal and professional) often go hand in hand.

Let's think about peacocks for a second. Peacocks have evolved great plumes of feathers for a very specific reason: to attract a mate. The more dazzling the plumage, the more desirable the bird. Nature abounds with examples of beautiful plumage, impressive musculature, or other physical traits that are desirable to members of the opposite sex, because these traits indicate reproductive fitness. For humans, a healthy, beautiful smile is one of those traits. Good teeth are an excellent indicator of desirability. They indicate good health and a commitment to self-care.

Broad U-shaped jaws that allow for optimal airways also make the smile broader. They provide enough room for all the teeth, which helps with the fullness of the lips. A correct bite leads to better chewing and less dental wear, but also results in facial balance and a proportional profile with the nose, lips, and chin in harmony. Nose-breathing allows for the optimal growth of the mouth and the face.

This may all seem like an overstatement—after all, looks depend on more than just the mouth, right? Amazingly, in many ways, they don't.

Remember that the maxilla (the upper jawbone, the middle of the face) and the mandible (the lower jawbone, the bottom third of the face) determine much of how the entire face grows and looks. Most of the face's structure depends on these two bones and how they affect the smile and the tissues around them: the lips, chin, nose, eyes, cheeks, and jawline. A skilled orthodontist has the knowledge and the tools to affect all of those facial features.

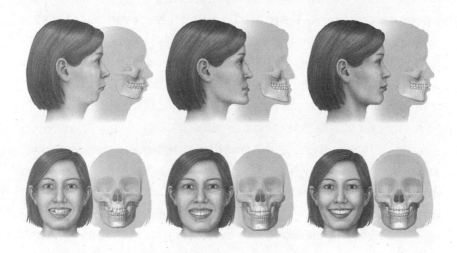

There is a great deal of shame around appearances in our culture and tremendous pressure to look good. This is especially true for women. In one survey, 85 percent of women and 79 percent of girls said that they opt out of important life activities when they don't feel good about the way they look.[8] And, sadly, 88 percent of teens feel self-conscious about their smiles.

The way we look can directly affect the way we experience life (and the way others treat us). Just look at the cult of celebrity. Besides being generally fit, coiffed, and well dressed, what is the one thing American celebrities have in common? Perfect teeth.

Whether we like it or not, physical looks affect our lives in profound ways. For example, we know that self-esteem and self-confidence are vital for a good life. Having a healthy, beautiful smile is key to building self-confidence. When someone feels confident, they are more likely to be perceived as happier, more attractive, more successful, and smarter.

Again, it's reasonable to argue that we should not be giving as much weight to the way people look—that we should evaluate people solely on their actions, personalities, skills, and other attributes. I wholeheartedly agree! But to avoid having this conversation because of those objections ignores reality. We live in a society obsessed with beauty. Because of that, attractiveness matters for a person's success in life. As an orthodontist, I have seen the profound transformation over and over—from self-conscious teen to self-confident adult. In so many cases, it all hinges on a healthy mouth.

What do you think is the number one feature people choose when you ask them what makes someone attractive? That's right, their smile.[9] Smile ranks as the most important physical trait when it comes to attraction—above all other factors.

Attractiveness plays a role in everything from finding a partner to landing a job. Often, it comes down to that great first impression. You have 100 milliseconds to make a first impression, and first impressions tend to become the lasting default view we have of people, even when we've known them for a while. Ninety-six percent of adults believe an attractive smile makes a person more appealing to members of the opposite sex.[10]

An American Association of Orthodontists report confirms that bad teeth represent the biggest dating turnoff among men and women. When considering a potential love interest, 77 percent of women think crooked teeth are worse than a receding hairline.[11]

When it comes to employment, good oral care is just as crucial. Americans perceive those with straight teeth to be 45 percent more likely to get a job than someone else who is evenly matched in skills and experience. A candidate's appearance ranks as more important to getting a job than where the candidate went to school.[12] Is this fair? Of course not! But it is the world we live in.

If you're not happy with your smile, you're not alone. Only 28 percent of Americans are.[13] Many people avoid smiling because they're self-conscious about their teeth. That means they're not enjoying all the advantages that smiling provides. These people may be missing out on mental health gains, positive relationships, employment opportunities, and other benefits because they don't feel good about their smiles. Think about what not smiling says about them to others. Someone who never smiles seems angry or mean, uninterested, or judgmental. Not smiling doesn't say, "I'm self-conscious about my mouth." It says, "I don't have time for you."

The good news is that we can prevent and treat oral and facial imperfections to boost physical, emotional, and mental health.

CONNECTING THE DOTS

Many of us go to a dental health professional only when we experience tooth pain or to straighten our teeth. But, in fact, an orthodontist can do so much more to improve the functionality and confidence of our smiles. Here's a surprising list of issues your orthodontist can fix:

- Airway obstruction
- Mouth-breathing
- Sleep-disordered breathing
- Tongue-tie or lip-tie
- Tongue thrust
- Thumb-sucking or other bad habits
- Skeletal discrepancies (underbite, overbite, cross bite, overjet)
- Facial growth issues such as long face syndrome
- Width of the smile (too narrow or too wide)
- Lip incompetence (lips not coming together when relaxed)
- Lip fullness (too flat or too protruded)

- Profile (balance and facial disharmony)
- Facial height and imbalance among parts of the face
- Clenching or grinding of the teeth
- Speech issues
- Dry mouth
- Bad breath
- Gummy smile

CHAPTER 5

ORAL HEALTH AND PHYSICAL DISEASE

IN THIS CHAPTER

- How oral health affects (and is in turn affected by) whole-body health.
- How dozens of diseases are connected to the mouth, from Alzheimer's to cancer.
- The oral care products you use could be doing more harm than good.
- A comprehensive list of ingredients you should look for in oral care products (and ones to avoid).

I remember my grandmother taking her dentures out before bed. She'd put them in a cup on the nightstand. It was just . . . what my grandmother did. I figured that's what all grandparents did. But as I got older and realized my grandmother was actually quite a bit younger than my friends' grandparents—that she was, in fact, not old at all—and that other people her age seemed to have all their teeth, I got curious. One day, when I was about 12, I asked her about it. With tears in her eyes, she told me the story.

When she was in her 20s, she had a full set of beautiful teeth. She was young and strong. But she started having some problems with her digestion, so she went to see her doctor. He had no idea what was wrong with her. She went back many times and tried many different treatments, but none of them worked. Finally, her doctors suggested the problem could be with her teeth. The solution they proposed was a total extraction. If they took out all her teeth, they posited, the digestive problems would go away.

So, that's what they did. It was horribly traumatic for my grandmother—a young woman, now without teeth. It was also traumatic for her family, as she struggled to adjust to her new circumstance. Just imagine how it would feel to be young and beautiful—in the prime of your life—and to suddenly have all of your teeth removed! How shocking must that have been for her? Even now, many years later, I shudder to think. Sadly, they never did find out the cause of my grandmother's digestive troubles. But I am pretty confident it wasn't rooted in her teeth.

It's interesting how the pendulum swings on medical and dental opinion. It has gone from one extreme to the other: from removing all the teeth to cure systemic illness, to completely ignoring the teeth as a potential source of disease. Today, we've landed in a more objective and reasonable center, but we're still reeling a bit, as a discipline, from the vagaries of the past. We've seen this happen with many other treatments too. We used to remove tonsils and adenoids at the drop of a hat, anytime we saw issues with recurrent infections or breathing. Now, we take a more thoughtful approach, removing them only when it's absolutely necessary. We keep learning that there's still so much we don't know about the human body, and so we've become more cautious about removing structures unless we absolutely must.

ORAL HEALTH AND SYSTEMIC DISEASE (A VERY OLD STORY)

Throughout history, people have noted the connection between oral and physical health. As early as the 5th century BCE, the ancient Egyptians and Greeks were writing about the link between a healthy mouth and a healthy body.[1] These early writings weren't very scientific and were mainly based on intuition and controversial observations, but they were accurate, nonetheless.

Centuries later, in 1891, a microbiologist named W. D. Miller published a paper entitled "The Human Mouth as a Focus of Infection," arguably starting a new era known as the era of "focal infection." Miller blamed oral microorganisms or their products for a long list of diseases including arthritis, diphtheria, tuberculosis, syphilis, thrush, pulmonary diseases, cardiovascular diseases, gastric problems, brain abscesses, and even stupidity (which was considered an actual medical diagnosis in the 19th century).[2]

In 1911, following an article written a decade earlier called "Oral Sepsis as a Cause of Disease," the English physician William Hunter blamed poor dentistry (and the resulting oral sepsis) for causing most chronic diseases.[3] Addressing the medical students at McGill University in Montreal, he argued:

No one has probably had more reason than I have had to admire the sheer ingenuity and mechanical skill constantly displayed by the dental surgeon. And no one has had more reason to appreciate the ghastly tragedies of oral sepsis which his misplaced

*ingenuity so often carries in its train. Gold fillings, crowns and
bridges, fixed dentures, built on and about diseased tooth roots
form a veritable mausoleum over a mass of sepsis to which there
is no parallel in the whole realm of medicine and surgery. A
perfect gold trap of sepsis of which the patient is the proud owner,
and no persuasion will induce him to part with it, for it cost him
much money and it covers his black and decayed teeth.[4]*

In 1912, an American physician, Frank Billings, similarly introduced the concept of focal infection to American physicians.

They were all on the right track, of course. But in their enthusi-
asm for their cause, and perhaps due to the sometimes-adversarial
relationship between medicine and dentistry, many physicians
started recommending tooth extractions in favor of restorations.
They began to think of it as the only true cure for a variety of ail-
ments. This thinking resulted in a generation of toothless people in
their prime, like my grandmother.

Fortunately, the science eventually prevailed, and the 1930s and
1940s saw a steep decline in baseless extractions and a reemer-
gence of restorative dental procedures.[5] But scientists continued to
wonder about the link between oral infection, inflammation, and
overall health. It really didn't come back into the mainstream consciousness until the end of the 20th century. That's when a series
of well-researched studies appeared on the associations between
periodontal disease and systemic conditions like coronary heart
disease and preterm birth. It finally became very clear why elimi-
nating dental infections or periodontal disease could help people
live longer, healthier lives.

In 2011, in response to growing global concerns over chronic
conditions like heart disease and diabetes, the United Nations (UN),
for the first time, officially associated oral diseases with significant
morbidity.[6] This high-level UN meeting was historic because it was

only the second time in history that a health topic was discussed at the United Nations by leaders of countries around the world (the first time was a summit on HIV/AIDS in 2001).[7]

Over the last 10–15 years, there has been a flood of scientific data supporting the idea that oral health dramatically impacts physical health. The case is so strong, and so different from what's been taught in the past, that many medical textbooks need to be rewritten in significant ways. In fact, the science in this area is growing and changing so much that a whole new branch of periodontology has emerged. It's called "Periodontal Medicine" and its entire focus is on the strong bidirectional relationship between periodontal disease and systemic health.

It's mind-boggling how many systemic diseases are now linked to periodontal disease. The list includes cardiovascular disease, diabetes, atherosclerosis, rheumatoid arthritis, pneumonia, adverse pregnancy outcomes, chronic kidney disease, erectile dysfunction, Alzheimer's, and many types of cancer. The relationship is real, but as with every relationship, it's complicated!

Systemic Diseases Linked to Periodontal Disease

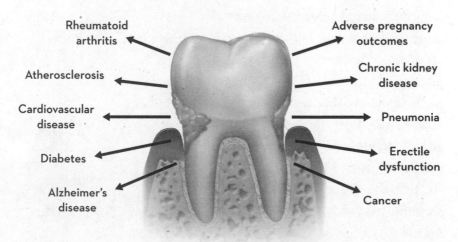

THE CONNECTION BETWEEN THE MOUTH AND THE BODY

Sometimes the relationship between oral health and systemic disease is unidirectional (one way): the periodontal disease (gum disease) causes the systemic illness directly. Sometimes, it's bidirectional (both ways): the periodontal disease contributes to the systemic disease, and the systemic disease contributes to the periodontal disease. Sometimes, both things have a common underlying cause.

To help you understand how your mouth's periodontal health is related to the rest of your body, let's follow the mouth's journey from healthy to mildly unhealthy (gingivitis or gum inflammation) to severely unhealthy (periodontitis).

In a healthy mouth, there is a delicate balance of microbes. It's in homeostasis—a healthy, balanced state. But when this balance is upset, opportunistic overgrowth can occur. This is often what starts the process of disease and decay. As a healthy mouth begins to get less healthy, gingivitis develops. The gums may bleed and look red. There may be bad breath or sensitive, swollen gums. Because gingivitis is usually painless, it can go for years before it gets diagnosed.

Gingivitis is reversible with good oral care but, left untreated, it often progresses to periodontitis. Periodontitis is a serious gum infection that results in the irreversible loss of the supporting structures of the tooth—the periodontium. These structures include the gums, alveolar bone, cementum, and the periodontal ligament. As periodontitis advances, teeth can get loose and

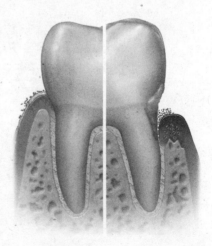

eventually fall out. As this process unfolds, bacteria can enter the bloodstream, and untreated infection can lead to chronic inflammation throughout the body. This is a bad situation for systemic health.

The mouth is filled with billions of microbes. Current estimates state that the oral microbiome contains about 700–1,000 species, although some estimates are as high as 1,200 or more.[8] The oral microbiome is not uniform—it has its own microhabitats in different areas of the mouth. You'll find different types of bacteria thriving on the tongue, palate, cheeks, teeth, and inside the periodontal pockets between teeth and gums that deepen as the periodontal disease advances.

Unlike other parts of the mouth and gut where the outer layer of tissue is constantly shedding and renewing, the microbes below the gum line don't get regularly flushed out or disrupted. They organize into colonies, form biofilms that coat teeth, and live there for long periods of time.

According to the Centers for Disease Control and Prevention (CDC), half of American adults aged 30 and over have periodontal disease, and the rate increases to over 70 percent for adults aged 65 and older.

RISK FACTORS FOR DEVELOPING OR WORSENING PERIODONTAL DISEASE

- Poor oral hygiene
- Poor diet
- Smoking/chewing tobacco
- Dental phobia (which leads many people to avoid the dentist)
- Crooked teeth (hard to keep clean, which increases the risk of damage to the periodontium)

- Poor bite (places uneven or excessive pressure on teeth and surrounding tissues)
- Poor oral habits
- A history of antibiotic use
- Harsh oral care products (certain ingredients such as alcohol can change the microbiome, cause dry mouth, even worsening bad breath)
- Stress
- Immunodeficiencies
- Defective fillings
- Medication side effects (steroids, seizure drugs, cancer drugs, drugs that cause dry mouth, and calcium-channel blockers that cause hyperplasia)
- Genetic factors (disease-susceptible individuals)
- Age
- Hormonal changes during pregnancy or with oral contraceptives, puberty, menstruation, menopause, and post-menopause
- Certain systemic conditions such as diabetes or autoimmune disease (like rheumatoid arthritis or lupus) can affect or interfere with the body's inflammatory system and can worsen the condition of your gums

CAN YOU CATCH CAVITIES AND PERIODONTAL DISEASE?

The simple answer? Yes.

But it's more complicated than that. (Surprise!)

Oral microbes can be transferred between people from kissing, sharing utensils, licking, biting—any transfer of saliva can do it. Parents commonly transfer their oral microbes to their children, though kids also get their microbiomes from other caregivers, siblings, friends, and even pets! Depending on the quality of the microbiome, this can be a good thing or a bad thing.

Where Do Those Microbes Go and What Are They Doing There?
So glad you asked!

There are lots of places microbes can go. Once they gain access to the bloodstream, they can travel anywhere in the body. For example, bacteria entering through bleeding gums or open root canals may travel to the heart, lungs, brain, liver, pancreas, GI system, joints, or to an unborn baby. Having bacteria in the bloodstream is very bad. It's called "bacteremia." If it's short-lived, it may cause no symptoms at all. But if bacteria accumulate in tissues or organs, they can cause life-threatening infections. One dangerous potential infection is called "endocarditis"—an infection of the lining of the heart. Generally, if enough bacteria build up in the blood, sepsis may occur. This is a whole body infection and is very serious indeed. These are the direct ways oral bacteria can cause systemic disease.

Stunning research published in 2019 points to a causative connection between a common oral bacteria and Alzheimer's disease.[9] Since 2016, researchers have known that amyloid plaques—the sticky substances present in the brains of Alzheimer's sufferers—seem to be one of the body's responses to bacteria. In the presence of bacteria, these plaques develop around bacterial cells, as a sort of defense mechanism. Scientists have also known that the presence of *Porphyromonas gingivalis*, a common bacteria involved in gum disease, is a risk factor for Alzheimer's. This bacterium has been found in the brains of deceased Alzheimer's patients in the presence of amyloid plaques, but it was unclear if the bacteria was a cause, or a result, of Alzheimer's.

In this most recent study, researchers infected mice with *P. gingivalis*, and then studied their brains. In these mice, gum infection "led to brain infection, amyloid production, tangles of tau protein, and neural damage in the regions and nerves normally affected by Alzheimer's. This suggests causation."[10]

If these results are replicated and we determine that gum disease can cause Alzheimer's, I predict a lot more attention will be paid to gum disease prevention, and, hopefully, to oral health in general. This feels like a potential public health eureka moment!

Bacterial infection in the mouth can also indirectly cause chronic whole body inflammation, and that can be very dangerous too.

INFLAMMATION AND DISEASE

If you've ever sprained your ankle, gotten a splinter, or been stung by a bee, you know about inflammation. It's that hot, red, swollen thing your body does in response to injury, and it's a vital part of staying healthy. When your body is injured, it responds in force, dispatching white blood cells and fluid to repair cells or fight off hostile microbes.

This is great news when the injury is acute and short term. As an infected wound heals, the inflammatory response subsides. The body has done its urgent triage, and you can go on your merry way. But, if the threat remains—as it does with chronic periodontal disease, for example—the inflammation does too. Those cells and chemicals meant to heal begin to circulate through the body. This is called "chronic inflammation," and it can cause grave damage.

A condition in which the immune system is perpetually "on" may not sound so bad. If the immune system is on all the time, you're extra safe, right? Unfortunately, the immune system doesn't work that way. A lingering immune response can begin to attack healthy tissues or organs.

Chronic inflammation is connected to many chronic diseases like Alzheimer's, diabetes, rheumatoid arthritis, heart disease, and more. Chronic inflammation can happen for many reasons, but one common cause is untreated gum disease.

ENOUGH ABOUT BACTERIA . . . WHAT ABOUT VIRUSES?

Good question. The oral microbiome isn't just composed of bacteria. Viruses are also common in the mouth, and they can also cause many different types of systemic problems.

HPV

One of the most common viruses in the oral microbiome is the human papillomavirus (HPV). The Cleveland Clinic reports that approximately 79 million Americans are currently infected with HPV, with an estimated 80 percent of sexually active people harboring it at some point in their lives. Even though HPV is the most common viral STI in America, the vast majority of cases go unnoticed and become undetectable after a year.[11] HPV may be transmitted through sexual contact or via kissing. In addition to causing most cervical cancers as well as some other cancers, HPV can cause genital warts. The CDC recommends that all children get the HPV vaccine at age 11 or 12.

Herpes

Herpes simplex virus (HSV) is another common visitor of the oral cavity and can spread easily through shared saliva. There are two types of herpes: Type 1 is the most common and is usually acquired in childhood. It can cause sores around the mouth and lips, called fever blisters or cold sores. Type 2 is usually spread through sexual contact and can cause sores around the genitals. Genital herpes can be passed on to the baby during childbirth, if the mother has an active infection.

Epstein-Barr Virus

Epstein-Barr virus (EBV) is one of the most common viruses in humans and can spread through bodily fluids, primarily saliva.

Many people become infected with EBV in childhood without serious symptoms, but the virus lingers in the body indefinitely. EBV can cause infectious mononucleosis (aka, the kissing disease, or mono) if infection occurs during adolescence. Ninety percent of adults worldwide are EBV seropositive.

BRUSH AND FLOSS TO SAVE YOUR ~~TEETH~~ LIFE

Without a healthy mouth, you can't have a healthy body. It's time we all took that to heart.

Evidence of the links between periodontal disease and many systemic diseases and conditions is growing stronger by the day. Here's a brief rundown of the connection between an unhealthy mouth and physical disease:

Alzheimer's Disease

In addition to the study mentioned earlier, another recent study determined that people who have had periodontitis for more than 10 years are 70 percent more likely to develop Alzheimer's disease than people without periodontitis, even after accounting for other factors such as the environment, diabetes, and heart disease.[12] Additionally, gum disease can speed up mental decline by six times in Alzheimer's patients.

Asthma

A 2018 review in the *Journal of Periodontology* of 21 studies conducted and published between 1979 and 2017 strongly suggests the association of asthma with periodontal disease.[13]

Cancer

Some types of cancer do run in families, so they may have a genetic component, but many cancers are linked to behavior and environmental factors such as smoking, diet, or inflammatory processes and infections.[14] The World Health Organization (WHO) has designated three common oral microbes as human carcinogens: Helicobacter or *H. pylori* bacteria, and HPV and EBV viruses. *H. pylori* has been associated with cancerous lesions of the upper gastroesophageal tract, including the mouth, and can also increase the risk of developing cancer of the colon or rectum by 50 percent. EBV has been associated with cancers including breast, kidney, oral, bladder, and thyroid. HPV has been blamed for 4.8 percent of all new cancers worldwide. HPV infections can be linked to cervical cancer (in almost 100 percent of cases), genital cancers, and oropharyngeal cancer.[15]

In a longitudinal study of 11,328 adults aged 25 to 74, those with periodontitis had a 73 percent higher risk of dying of lung cancer than those with healthy gums, even after accounting for other variables such as smoking.[16]

A 2007 study from Harvard University demonstrated a striking correlation between pancreatic cancer and gum disease in men.[17] Those with gum disease had a 63 percent higher incidence of the deadly cancer.[18] Another study by a team from both Harvard and Imperial College London looked at health records from 50,000 patients over 21 years of age; it found that gum disease correlated with a significantly higher risk of several different cancers, including lung (33 percent), kidney (50 percent), and blood (30 percent).

When gum disease was chronic and advanced, head and neck cancer risk increased fourfold for each millimeter of bone loss around teeth.[19] According to recent research from the American Association for Cancer Research, women who have gum disease are more likely to get breast cancer. Women who reported a history of gum disease in the study had a 14 percent increased risk of cancer overall.

Cardiovascular Disease

Inflammation appears to be behind the relationship between peri-odontal disease and cardiovascular disease (aka heart disease).[20] Cardiovascular disease occurs when plaque (a completely different type of plaque from dental plaque) builds up in the walls of arteries, causing a narrowing of those arteries (atherosclerosis). Atherosclerosis makes it more difficult for blood to flow through the arteries, raising blood pressure. If a blood clot forms in the arteries, it can stop the blood flow entirely, leading to a heart attack or stroke.

One landmark study showed that bacteria from the mouth and gut can be the source of atherosclerotic plaque-associated bacteria.[21] Another study showed that you are 4.3 percent more likely to suffer from cardiovascular disease if you have poor oral hygiene and 19 percent more likely if you have periodontal disease.[22] Other researchers have shown that after adjusting for risk factors, such as smoking, alcohol, obesity, blood pressure, and diabetes, patients with periodontal disease had a 1.14-fold greater risk of developing cardiovascular disease than those without periodontal disease.[23]

Chronic Kidney Disease

In a study of 5,500 patients, those with periodontal disease were much more likely to have renal insufficiency—a condition that slowly reduces kidney function leading to potentially life-threatening renal failure.[24] In another study, patients with chronic kidney disease who also had periodontitis showed an increased risk of death from those with healthy gums.[25]

Diabetes

According to the World Health Organization, there are about 347 million adults worldwide with Type 2 diabetes, and this number is expected to double by the year 2030. Diabetes is increasingly common in adults today, with almost one in every ten adults (9.4 percent of the population) now suffering from the disease.

Diabetes and periodontal disease have a bidirectional relationship. Remember that "bidirectional" in this context means the periodontal disease contributes to the systemic disease, and the systemic disease contributes to the periodontal disease. Diabetes increases the risk of periodontal disease because people with diabetes are more susceptible to contracting infections.

Then, in turn, periodontal disease makes it harder to control blood sugar in diabetics.[26] As you may imagine, overall outcomes are worse when these two conditions are combined. People with diabetes who also have periodontal disease are about three times more likely to experience complications such as kidney disease. They are also more likely to get hospitalized for diabetes. What is perhaps scariest of all is that nearly half of adults suffering from Type 2 diabetes are not even aware they have it.[27]

Endocarditis

As I mentioned earlier, infective endocarditis is a devastating disease in which the inner layer of the heart, the endocardium, becomes inflamed. The heart valves are usually involved, although other surfaces and structures can be affected as well. A 2021 study published in *Microorganisms* states that excessive oral plaque can increase your risk of infective endocarditis-related bacteremia, which globally affects 7.46 million people. Endocarditis has a high mortality rate of 20 percent.

Erectile Dysfunction

Periodontal disease is a potential risk factor in the development of erectile dysfunction. A review in the *Journal of Reproductive Science* of 10 international studies, published between 2009 and 2016, showed ED to be more than two times more common for men with periodontitis than men without it, even after accounting for diabetes.[28]

High Cholesterol
High cholesterol is a risk factor for heart disease, heart attacks, and stroke. It turns out periodontal disease is associated with elevated plasma triglycerides and total cholesterol. Periodontal therapy reduces arterial hardening in much the same way as a 30 percent drop in LDL (bad) cholesterol.[29]

Hypertension
High blood pressure is a major risk factor for cardiovascular disease. In the U.S., high blood pressure affects almost 30 percent of the adult population aged over 18. No surprise here, high blood pressure is associated with periodontal disease. A thorough review of the current research, published in 2020 in *Cardiovascular Research*, showed that the more severe the periodontal disease, the higher the risk of hypertension. There is now evidence that treatment of periodontal disease may significantly decrease blood pressure.

Inflammatory Bowel Disease and Liver Disease
Inflammatory Bowel Disease includes several conditions such as Crohn's disease and ulcerative colitis. IBD is also associated with periodontitis.[30] In one study, patients with IBD had an increased risk of having periodontal disease, especially those with Crohn's disease.[31]

Periodontitis promotes the progression of fatty liver disease, which can lead to cirrhosis and liver failure.[32]

Obesity
Obesity has long been considered a risk factor for many systemic conditions such as cardiovascular disease, diabetes, sleep apnea, and arthritis. There is an increasing body of evidence establishing

a relationship between obesity and periodontal disease. A study from the journal *Oral Diseases* found a sixfold-higher risk of severe gum disease in overweight people. In recent years, there has also been a growing number of studies on the possible link and potential contributing factors between the human gut microbiome, the oral microbiome, and obesity.[33]

Osteoporosis

In a review of 17 studies published between 1998 and 2010, the majority showed a positive relationship between osteoporosis and periodontal disease. This indicates that osteoporosis should be considered a risk factor for periodontal disease progression. In addition, it's important to note that the jaws of the mouth are bones like the other bones in the body, and so the same complications (disease, medication, or aging) that affect any bone in the body can impact the jaws.[34]

Pregnancy Complications

As we discussed in Chapter 2, periodontitis is associated with infertility, premature birth, low birth weight, and other pregnancy complications.

Respiratory Diseases: Pneumonia, Acute Bronchitis, COPD, and COVID-19

When a person has poor oral hygiene and periodontal disease, there is a risk that they will inhale the opportunistic bacteria from the mouth and upper throat into the lower respiratory tract. This can cause debilitating and even life-threatening respiratory infections such as pneumonia, acute bronchitis, or chronic obstructive pulmonary disease (COPD). This is not surprising, since the surfaces of the mouth are contiguous with those of the trachea and lower airway. Several studies have shown that periodontal disease is associated with the increased prevalence of respiratory diseases, even after adjusting for confounding factors.[35]

A 2021 study published in the *Journal of Clinical Periodontology* suggested that patients with gum disease were nine times more likely to die from COVID-19. They also found that people with periodontal disease were at least three times more likely to be admitted to the ICU and 4.5 times more likely to require a ventilator. In this latest study, inflammation seemed to explain the raised COVID-19 complication rates.[36]

Rheumatoid Arthritis

Rheumatoid arthritis (RA) is a chronic inflammatory disorder affecting the joints, such as those in the hands and feet. Periodontal disease is highly prevalent in patients with RA. Research suggests that taking care of your teeth may be a good way to take care of your joints.[37] According to several studies, tooth loss as a marker for periodontal disease may predict rheumatoid arthritis and its severity. In other words, the more teeth lost, the greater the risk of RA.

Stroke

Several studies have concluded that periodontal disease is a risk factor for stroke. In one of the largest studies of its kind, researchers showed a graded relationship between the extent of periodontal disease and stroke risk. Additionally, they found that regular dental care was associated with cutting the stroke risk in half.[38]

I could go on and on and on. Since periodontal disease is linked to so many systemic health problems, it's likely that this link extends to many others that we haven't even discovered yet.

LET'S PUT IT IN PERSPECTIVE

When you look at the relationship between periodontal disease and mortality, there is a deeply disturbing correlation. People with gingivitis and periodontitis have a 23 to 46 percent higher chance of dying than those who are free of this disease.[39] Young men under age 50 with advanced periodontal disease are two and a half times more likely to die prematurely and three times more likely to die from heart disease than those with healthy mouths.[40] In fact, researchers have shown that the more missing teeth a person has, the poorer their quality of life, and the higher their risk of premature death. Elderly individuals without any teeth have a 30 percent higher risk of death compared to those with 20 teeth or more.[41]

Your daily oral hygiene habits affect your longevity too. Here are a few more stats to chew on:

- Never brushing at night increases mortality risk by 20–35 percent as compared with nightly brushing.
- Never flossing increases mortality risk by 30 percent compared with flossing every day.
- Not visiting a dentist within the last 12 months increases mortality risk by 30–50 percent as compared with seeing a dentist two or more times per year.

Percent Increase in Mortality Risk

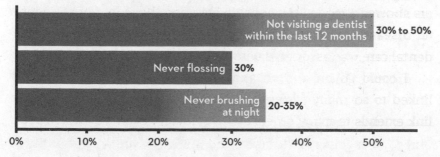

These stats are shocking. But I hope they are starting to make sense. Your mouth supports your good health and can cause your bad health. Taking care of it is one of the best things you can do to protect your quality of life and longevity.

It's important to remember that there are many compounding factors when it comes to health. But when you add periodontal disease to the mix, all of the negative outcomes become much more likely.

When the solution is so easy—brush, floss, visit your dentist regularly—it becomes crystal clear that the way many of us overlook oral health must change.

YOUR MOUTH: A MIRROR FOR YOUR BODY

Just as your mouth can cause illness in your body, it can reflect your body's illness. Early signs of many medical problems appear in your mouth, and early diagnosis and management can be very helpful in treatment outcomes.

For example, in addition to the usual suspects, gingivitis can be caused by serious conditions such as leukemia or HIV. Oral ulcers can be indicative of viral or fungal infections. Oral candidiasis (thrush), which is the most common fungal infection, can be caused when the normal oral microbiome is altered, but can also be a sign of immune dysfunction. Dry mouth, canker sores, and changes in the gingiva can be caused by many medications used to treat a variety of disorders.

It is estimated that more than 100 systemic conditions and hundreds of medications can have oral manifestations. Here is a brief summary of the oral manifestations of some systemic diseases:[42]

Gastrointestinal Diseases

Since the esophagus, stomach, and intestines are all part of the GI tract—and the mouth is the port of entry to the whole system—it follows that the mouth is often involved in GI conditions.

- **Gastroesophageal Reflux Disease (GERD)**
  Although occasional reflux can happen in healthy people,
  when reflux is recurrent and the acid is causing damage
  to the esophagus or the mouth, it gets classified as GERD.
  It is estimated that 20 percent of Americans have GERD.
  In addition to heartburn, which is a common symptom,
  gastric acid can cause damage to the teeth and soft tissues
  of the mouth, cause changes in saliva, and cause halitosis
  (bad breath). In addition to GERD, dental erosion (wear of
  teeth) can also be seen in patients with eating disorders
  such as bulimia.

- **Crohn's Disease**
  Crohn's disease is an inflammatory bowel disorder (IBD)
  that involves the lining of the digestive tract. It can lead
  to stomach pain, severe diarrhea, fatigue, weight loss,
  and poor nutrition. Up to 30 percent of pediatric patients
  with Crohn's disease may show early signs in the mouth
  even before intestinal involvement begins.[43] These signs
  include gingival swelling of the gums and mucosa, and
  mouth ulcers or fissures at the angles of the mouth.
  Impaired speech and malnutrition can follow, since
  patients often have difficulty eating and speaking because
  they are in pain.

Chronic Liver Disease (CLD)

The liver is involved in many body functions, such as filtering the
blood, detoxifying chemicals, and metabolizing drugs. CLD affects
approximately 15 percent of the U.S. population and refers to various diseases of the liver that last more than six months. Oral signs
can include excessively bleeding gums and jaundice (a yellow tint

in the thinner areas of the mouth, such as the soft palate and under the tongue).

Thyroid Disorders

Thyroid hormones are essential for regulating metabolism. The American Thyroid Association warns that thyroid disorders affect up to 12 percent of people and can be due to either over-functioning (hyperthyroidism) or under-functioning (hypothyroidism) of the thyroid gland. Oral manifestations are more common in severe hypothyroidism and can include swelling of the lips and tongue, delayed dental eruption, and macroglossia (large tongue) if the condition is congenital. Patients with hyperthyroidism can be more susceptible to developing cavities and periodontal disease too.

Cancers

- **Leukemia**
  Leukemia is a cancer of white blood cells. Gingival hypertrophy and hyperplasia (overgrowth of gum tissue) are common signs of acute leukemias. Also, oral ulcers can be caused by the chemotherapy drugs involved in the treatment of leukemia.

- **Lymphoma**
  Cancers of lymphoid tissue and lymphoid precursors are the seventh most common cancer in the U.S. Lymphomas in the mouth are generally painless, soft masses that can also accompany ulcers.

Nutritional Deficiencies

Deficiencies in vitamins and minerals can be caused by poor diet, but they can also be a result of poor absorption related to a variety of systemic conditions. These nutritional deficiencies can have many signs and symptoms in the mouth ranging from taste changes to a burning tongue, and from gingival bleeding to loosening or lost teeth.

TREATMENT OPTIONS FOR PERIODONTAL DISEASE

As always, prevention is best. Early gingivitis is easily reversible with daily brushing and flossing, using the right oral care products (see the end of this chapter for recommendations), and visiting the dentist regularly for a professional cleaning. Once the disease advances into periodontitis, treatment options become more advanced as well.

Initially, nonsurgical treatment options are available such as scaling and root planing, where your dentist will remove the tartar off your teeth and root surfaces. As the disease advances further, surgical options may become necessary, such as pocket reduction procedures, gum grafts, laser treatments, and regenerative procedures. Your dentist may also recommend local or systemic medications to treat advanced periodontal disease.

THE ECOSYSTEM WITHIN

Think of your mouth as an ecosystem—a natural community of living things—because the truth is, the communities of microorganisms in a human mouth have evolved, just as we have, over millions of years. Just like any habitat, the ideal state for this ecosystem within is harmony and balance. A balanced ecosystem is one in which populations keep each other in check. Imagine a forest where plants, insects, and animals compete for limited resources and have to adapt to the unique environmental challenges of their particular habitat.

When it comes to your mouth, its own habitat, you want that same competition for resources, because that keeps your oral microbes in balance. You want a healthy, self-sustaining oral ecosystem. This is your mouth's ideal state.

For many years, people have been overzealously using harsh foaming toothpastes, antiseptic mouthwashes, and antibiotics to wipe out the bacteria that cause cavities or gum disease. This is common practice, even in the absence of any oral disease. But what do you think that indiscriminate attack does to the delicate balance of microbes in the mouth? What it's *not* doing is keeping populations in check, promoting the balance that took millions of years to evolve, that's for sure.

Antiseptic mouthwash kills 99 percent of everything, as advertised on the bottle. What does it leave behind? The baddest, toughest, roughest little microbes around—poised to recolonize that entire mouth, totally unchecked by the organisms that used to hold them at bay. And, indeed, it's the same thing that happens with other antibacterial toothpastes and antibiotic treatments.

In addition to disrupting balance, antibiotics create the perfect conditions for antibiotic resistance. When that 1 percent of leftover microbes begin to reproduce and thrive, the next swish with mouthwash or brush with toothpaste isn't going to be nearly as effective. Eventually, you're stuck with a mouth full of super microbes immune to antibiotics.

As you can see, this "simple solution" is no solution at all.

The so-called oral care products we put in our mouths (and in the mouths of our children) can be dangerous in other ways as well. Mouthwashes and toothpastes have warning labels instructing us to get medical help or contact a poison control center if they're ingested. A 2018 FDA study showed that a very common toothpaste antibacterial ingredient, triclosan, can increase rates of colitis and colon cancer in mice. Another study the following year by the

American Dental Association linked triclosan to increased risk of osteoporosis in adult women. Triclosan was also found to be associated with a decrease in the levels of some thyroid hormones and making bacteria resistant to antibiotics. In 2017, the FDA restricted the use of triclosan in hand soaps and antiseptic sanitizers, and by early 2019, triclosan was no longer available commercially in toothpaste. But how many people routinely ingested this dangerous chemical for years before it was finally taken off the shelves? How many illnesses were caused or exacerbated by it?

Throughout recent history, human interference with many natural habitats and ecosystems has resulted in similarly negative unintended consequences. For example, in 1958, Chinese leaders led by chairman Mao Zedong launched the Great Leap Forward, an ambitious movement to rapidly improve the economy, health, and social conditions of China. One of the first actions taken was an initiative called the Four Pests Campaign. It was a project aimed at eradicating mosquitoes, flies, rats, and sparrows, the pests responsible for spreading disease and blighting crops countrywide.

Every man, woman, and child in China—an estimated 613 million people at the time—was called upon to do their patriotic duty. The campaign particularly targeted sparrows, since they were eating the grain seeds. Citizens were urged to frighten and shoot them—to do whatever it took to kill them off. Their nests were destroyed, and the eggs were crushed. People would bang on their pots and pans to scare them from resting on tree branches so they would fall to their death from exhaustion. The crusade against sparrows pushed them to near extinction. However, as the sparrow population declined, so did the crops. In 1960, scientists discovered that sparrows ate insects far more often than they ate grains, and with the decrease in sparrows came a huge increase in those insects. In the absence of predation, the locust population boomed, destroying the grain fields. The situation got so bad, the Chinese government began importing sparrows from the Soviet Union, but to no avail. The sparrow decline was partly responsible for the Great Chinese Famine when over 30 million people died of starvation.

This has happened over and over again throughout history. When human beings disrupt ecosystems, problems tend to follow. Yellowstone became a national park in 1872 and there were no protections for its wildlife. Indiscriminate hunting of wolves led to their local extinction by 1926. It was the wolves, however, that kept the elk population in check, and without them, elk numbers exploded. The elk overgrazed the land, causing massive problems with erosion, and became Yellowstone's number one ecological threat. Eventually, wolves were reintroduced, and they had a dramatic effect on the entire ecosystem. Fewer elk meant plants could recover, reducing erosion. Less competition for limited forage meant the beaver population increased, too, creating more dams, which improved the local watershed. Moose, mink, birds, fish, and amphibians all started doing better. Yellowstone still isn't back to its original glory, but it is a hopeful story, nonetheless.

In these examples, disastrous unintended consequences happened after we, humans, eliminated one species in these ecosystems. Imagine what could happen when we eliminate hundreds of microbe species in the ecosystems within our bodies.

It's both unfortunate and fortunate that maintaining your microbial balance is more complicated than brushing with a random toothpaste or rinsing with a random mouthwash. Unfortunate because you have to think bigger and do more. Fortunate because overall health is the answer, and that's good news for your long-term prospects. You need to take care of your mouth habitat: Give it good nutrients by eating well and avoiding the foods that feed the undesirables. Keep those nasties controlled by keeping your mouth reasonably clean—brush, floss, use only oral care products that are truly safe and effective in promoting oral health (see the end of this chapter for recommendations), and visit the dentist regularly. Occasionally, when necessary, seek out skilled interventions to address problems, much like a conservationist might intervene in a troubled forest, with the aim of reestablishing the natural balance.

Remember: you get only one mouth. If you think of it as an ecosystem—a balanced community—you're on your way to good oral health. If your mouth could talk, it would ask you to take good care of it and nurture your microbial communities to keep you healthy.

A healthy mouth leads to a healthy body, and vice versa. Care for one and you care for the other.

YOUR COMPLETE GUIDE TO ORAL CARE PRODUCTS

When I was growing up, I had a ton of cavities—a lot more than I would be willing to admit. And my mouth was so small that I had several of my teeth removed to make space when I got braces, which lasted more than three years. I even had to get a second set of braces during my orthodontic residency for two more years.

If it were all genetics, my son would be out of luck and doomed to have a mouth full of crooked teeth and cavities. Yet, he has no cavities and his bite and space for his teeth are as ideal as they can be. With a pediatric dentist mom and an orthodontist dad, we weren't going to let luck have anything to do with his oral health.

I've already shared with you all the pieces of the puzzle to ensure your children grow up with a healthy mouth. But I've left one final piece of that puzzle for last. And in some ways, it may be the most important piece.

Which oral care products you use at home and how you use them can have a profound impact on your health.

Aside from your dentist, you are your child's number one dental health guide. After all, you spend only two days out of the year at the dentist's office. That means how you and your children take care of your mouths the remaining 363 days can affect your health tremendously. In fact, using the wrong products or doing things the wrong way may even negate many of the positive steps you'll be taking to care for your mouth.

Nearly half of the world's population is suffering from oral diseases, including cavities and gum disease. And dental caries is the most common chronic disease in children. Clearly, what we are currently doing is not working. To turn things around, we need to rethink and do things quite differently to make sure we are getting the results we seek for our families.

YOUR ORAL CARE PRODUCTS MUST BE S.U.P.E.R.™

Every oral care product you're considering for your family must pass the S.U.P.E.R. checklist:

Safe
Unified
Playful
Effective
Reputable

Safe

Our environments are full of substances that can damage health—from pesticides and heavy metals in the food we eat, to car exhaust and other pollutants in the air we breathe, to toxic materials in cleaning products and furniture.

Environmental toxins affect babies even before they're born. In one study, 287 chemicals were detected in umbilical cord blood.[44]

Out of that huge number, 180 are known to cause cancer in humans or animals, 217 are toxic to the brain and nervous system, and 208 cause birth defects or abnormal development in animal tests.[45] The upshot: we are exposed to thousands of potentially dangerous chemicals.

Considering this, you'd think that the oral care products we put in our mouths and the mouths of our children on a daily basis would be safe, right? Unfortunately, many of the conventional oral care products contain, at the very least, questionable safety components, and some have downright toxic and dangerous ingredients. And don't be fooled by just any "natural" product, either. Many "natural" toothpastes simply replace certain toxic ingredients with others, and they still use ones that they shouldn't. For example, sodium lauryl sulfate (SLS) is a foaming ingredient often used in "natural" toothpastes that can, under certain circumstances, damage the soft tissues of the mouth, cause canker sores, and even be toxic to the body when used in large quantities.[46]

Ingredients from oral care products can be absorbed through the tissues and mucous membranes in the mouth and enter the bloodstream. From there, they can travel throughout the body. And what are the chances that your child swallows some of the toothpaste or mouthwash they use? Almost 100 percent. So, ingredients matter a lot.

Unified

You should always think of your oral care products as a system, or as a group of unified products that should work together and complement each other to accomplish your oral health goals.

Your mouth care products are only as good as the weakest link. For example, if you're doing everything right, but you're decimating your oral microbiome with that swish of fluorescent alcoholic mouthwash, you could be negating all your efforts.

Playful

Brush, floss, rinse, spit, repeat! Let's be honest, oral care is not something we typically associate with fun.

Many people dislike (okay, hate) the dentist and practice good oral care begrudgingly. They reluctantly brush and exaggerate about how much they floss to their dentists and do the absolute minimum just to get by. But what if people loved taking care of their mouths?

Our psychological associations with oral care are formed when we are young. Whether these associations are positive or negative depends on our early experiences. What would it be like to have your kids beg you to buy them their oral care products and use them enthusiastically and with pride? What if they were the oral health ambassadors in your family? They can and they should be.

Effective

Some products are safe but only partially effective (or even completely ineffective) in promoting oral health and treating disease. Of course, there are products that are unsafe *and* ineffective. For example, many of the so-called "natural" toothpastes or mouthwashes don't have any ingredients that effectively protect or support the mouth, while having antibacterial ingredients, such as potent essential oils, that can indiscriminately destroy the delicate balance of your oral microbiome. And, of course, oral care products should try to help support *all* parts of your mouth: teeth from outside and inside, the gums and bones, and your microbiome.

To make things worse, many oral care products are actually acidic. This either is because of their formulation or because their ingredients are designed to extend shelf life. Remember what we discussed in Chapter 1? What happens to a mouth microbiome when you introduce a lot of acid? Oral care products need to be pH balanced or alkalizing for maximum effectiveness.

Reputable

Your oral care products should be selected based on quality, science, and clinical research, and the results you and your children are looking for, rather than on the latest trend, or what some random person on social media or in the grocery store with little to no knowledge of dentistry recommends.

Are you getting the results you're looking for? Are your kids building positive associations with their oral care? Are their mouths healthier? Do they get fewer cavities? The answer to all these questions should be yes. If it isn't, you need a new set of oral care products for your family.

CHOOSING THE RIGHT PRODUCTS

Feeling confused about how to choose the right products for yourself and your child? Don't worry, I've done the work for you. Here is what you should look for in oral care products:

Toothpaste

To begin with, the reason it's called toothpaste is because many of the manufacturers treat the mouth like it's merely a collection of dead objects that need to be "zapped" with fluoride. They make toothpaste foam profusely to make you think it's actually doing something and give it a strong mint flavor that masks the bad breath caused by poor oral health. Some "conventional" toothpastes completely neglect the rest of the mouth, including the gums, and don't provide the nutrients that the mouth structures need. Last, but not least, many are out to kill all the microbes in your mouth, instead of trying to balance your microbiome.

The oral care market has recently been inundated with "natural" toothpastes, but many of them are not much better than conventional ones. In fact, not only do they lack effective ingredients to combat oral conditions like cavities or gum disease, they are also filled with ingredients such as potent essential oils that can decimate your oral microbiome and expose your mouth to all sorts of opportunistic microbes. Some also have additives that don't belong in the mouth at all. Additionally, they often use cheap, unhealthy sugars, such as sorbitol, saccharin, and aspartame, to improve their flavor.

Let's demystify toothpaste ingredients and show you what to look for and what to avoid.

THE FLUORIDE CONTROVERSY

Fluoride is present in some foods and naturally occurs in varying amounts in water sources such as lakes and rivers. It has also been added to drinking water in some parts of the country and is the active ingredient in many oral care products.

How does fluoride work to strengthen enamel?

When your saliva becomes acidic and the pH drops to about 5.5, small amounts of hydroxyapatite dissolve out of the enamel surfaces during demineralization. In the presence of fluoride, they combine to form a new mineral, fluorapatite. When pH rises again, the new fluorapatite minerals get deposited into the enamel surfaces during remineralization. Fluorapatite doesn't begin to dissolve from the teeth until the pH reaches about 4.5, so it is more resistant to acid attacks, making teeth stronger and less susceptible to cavities.

The American Dental Association recognizes the use of fluoride and community water fluoridation as safe and effective in preventing tooth decay for both children and adults.

If fluoride is effective in strengthening teeth, why all the controversy?

All toothpastes and mouthwashes containing fluoride are required by the FDA to have the following warning label on them:

Warning: Keep out of reach of children under 6 years of age.
If more than used for brushing is accidentally swallowed, get
medical help or contact a Poison Control Center right away.

It is true that fluoride, when applied topically, can help fight cavities. However, too much fluoride from any source (especially from dietary consumption) over a long period of time can potentially cause serious side effects. This is especially true for young children, since their teeth and bodies are still developing.

A recent CDC study revealed that nearly 40 percent of children use too much fluoride toothpaste when they brush. Young children also tend to swallow more toothpaste and mouthwash. The prevalence of dental fluorosis, a condition caused by childhood ingestion of fluoride and associated with tooth discoloration and abnormal enamel development, has become all too common. Additionally, if fluoride is swallowed in large amounts, it can potentially lead to serious toxicity and neurological problems.[47]

If you're considering using a fluoride toothpaste for your children, be sure to use no more than a tiny grain-of-rice-sized smear for children younger than three. For children between three and six years of age, the amount can be increased to a pea-sized dollop.

Hydroxyapatite

Hydroxyapatite is a calcium phosphate mineral that is naturally present in teeth and bones. In fact, it is the main substance in dentin (70 percent), and the enamel is almost entirely made up of hydroxyapatite (97 percent), which is the source of its strength and bright, shiny appearance. Since its invention as an ingredient in oral care by NASA in 1970s to help astronauts with their teeth remineralization, hydroxyapatite has been extensively studied and tested.

Although fluoride compounds are the only anti-cavity agents currently approved by the FDA, hydroxyapatite has been the gold standard in Japan for nearly 30 years and is also approved as a tool for cavity prevention in Canada. In fact, several studies have shown that, when used correctly, hydroxyapatite is safe and effective in remineralization of teeth.[48] This makes hydroxyapatite a great choice for children and a viable alternative for parents who want a fluoride-free option.

But hydroxyapatite (HA) does more than just help remineralize enamel. Studies have shown that hydroxyapatite can form a protective buffer on the enamel, lessening risk of erosion caused by sodas and other acidic beverages. And since hydroxyapatite mimics natural enamel, it can be effective in reducing tooth sensitivity by repairing enamel and exposed dentin, and it can naturally whiten teeth and restore luster damaged by bleaching agents or aging.[49]

Vitamin D and Vitamin K2

As you already know, your teeth are more than just inanimate objects that need repairing only from the outside. Teeth are very much alive on the inside, and calcium needs the help of two essential, fat-soluble nutrients, vitamins D and K2, to help support our teeth from the outside *and* inside.

Vitamin D, also known as the "sunshine vitamin," is a very important nutrient, and one of its key functions is to promote calcium absorption and to help carry it to our bones and teeth. Vitamin K has two main forms: K1 and K2. Vitamin K1 is primarily involved in blood clotting, while vitamin K2 serves to activate the proteins involved in getting calcium into the bones and teeth. Studies show deficiencies in vitamins D and K2 increase the risk of cavities and gum disease.[50, 51]

While we should be able to get adequate amounts of vitamin D by simply spending some time outdoors, the majority of people are deficient in it because of our modern lifestyles. Similarly, vitamin K2 is a powerful nutrient and though available in some animal and fermented foods (such as goose liver and natto), it's rare in the Western diet and is deficient in about 90 percent of the population. Most of us need help in getting adequate amounts of these vital nutrients into our bodies via supplements or oral care products.

Prebiotics

Earlier in the book, we compared the mouth to a garden, an interconnected ecosystem where all the different parts work together for the common good. As in any garden, you want to remove invasive weeds while keeping all the flowers healthy. Think of your toothpaste as a weed killer and fertilizer in one. You want it to get rid of the bad apples, while keeping the good ones healthy.

Unfortunately, many toothpastes act only like weed killer rather than fertilizer. They have one common goal: to indiscriminately kill all microbes that they come into contact with. To accomplish this slash-and-burn objective, some conventional toothpastes use strong antibacterial ingredients, while many natural toothpastes use herbal antimicrobials such as peppermint, tea tree, or eucalyptus essential oils, all of which can decimate the microbiome. These essential oils can be dangerous in other ways too.[52] For example, peppermint oil should never be used in young children since it can cause life-threatening breathing problems.[53]

We need to support our good microbes with nutrients, while selectively reducing the bad ones, and we can do just that with the right prebiotics. Prebiotics are powerful compounds that can feed and promote the healthy growth of beneficial microorganisms while starving the harmful ones. Three of my favorite prebiotics in oral care products are inulin, xylitol, and erythritol.

Inulin is a naturally occurring dietary fiber found in many plants such as chicory roots. In addition to keeping your mouth healthy, inulin can eliminate bad breath by encouraging the growth of good bacteria and inhibiting the obligate anaerobes associated with malodor.[54]

Xylitol is a type of naturally occurring sweetener found in many fruits and vegetables. It is called sugar alcohol, although it contains neither sugar nor alcohol. Xylitol is found in some oral care products because of its great taste and its ability to boost oral health and help prevent tooth decay.

Harmful oral bacteria feed on sugar from food, but they cannot metabolize xylitol. In the presence of xylitol, the bacteria ingest it and are unable to take up sugar and are effectively starved to death. In a systemic review of hundreds of studies between 2000 and 2019, xylitol consumption was found to reduce cavity-causing bacteria without changing the overall microbial composition and was thus determined to have properties of an oral prebiotic.[55] Xylitol can also help repair enamel damage by stimulating saliva and increasing the pH.

I should note that xylitol is not as easy to find in over-the-counter products as other cheaper and less effective sweeteners like sorbitol. But cariogenic bacteria (the ones that cause cavities) can still ferment sorbitol to some degree, so it's worth seeking out a product that contains xylitol, preferably one with at least 10 percent concentration for optimal effectiveness.

Erythritol is another sugar alcohol that tastes great and also has a cooling effect on the tongue. It is a noncaloric sweetener that shares many of the functional properties of xylitol. Although less studied than the more common and well-researched xylitol, some evidence suggests erythritol may have better efficacy compared with other sugar alcohols to maintain and improve oral health.[56]

What about probiotics?

Probiotics are live strains of bacteria that may be introduced in oral care products to add to the population of good microbes in your mouth. Although probiotics will perhaps have a promising role in the future, the current research has focused on specific strains in isolation, which may not represent their impact within a variety of oral microbiome mixes in different individuals. Additionally, unless the pH balance is managed, it is unlikely that probiotics will have a sustainable effect on the microbiome makeup. Therefore, I prefer to use products containing prebiotics that feed and stimulate growth among the preexisting good bacteria in every individual's unique oral habitat.

CHEW ON THIS

By now you might be thinking: "How do I choose which toothpaste is right for my kids? Do they need more fluoride or hydroxyapatite?"

The great news is that these ingredients are not mutually exclusive. You don't have to choose just fluoride or just hydroxyapatite. It's not an either/or scenario. They can all benefit oral health, alone or in combination. For example, a young child who doesn't have access to acidic or sugary foods (and therefore has a lower risk of cavities) could use a toothpaste with hydroxyapatite and prebiotics. On the other hand, an older person with a history of cavities and a daily soda habit might choose to add fluoride to the mix of hydroxyapatite and prebiotics to provide extra protection.

Please refer to the end of the chapter for a summary of all the ingredients you and your children should avoid and the ones you should always use.

Mouthwash

I wish I could say most over-the-counter mouthwashes are just a waste of money, but I can't, because they're even worse. If you're using a strong mouthwash daily to mask your bad breath, it may be causing you more harm than good. The strong menthol-flavored, neon-colored, alcohol-based mouthwashes can wreak havoc on your oral microbiome and dry your mouth, which can actually worsen bad breath.

But, of course, as with everything else, not all mouthwashes are created equal. With the right mix of ingredients, mouthwash can be a wonderful addition to the overall oral care system. A good mouthwash can loosen food particles, reduce plaque, and provide the right ingredients to the parts of your mouth not easily accessed by a toothbrush. Mouthwash containing alkalizing ingredients, such as sodium bicarbonate (baking soda), can also help keep the pH of the mouth balanced to promote a healthy oral microbiome.

Many of the toothpaste ingredients I discussed earlier also apply to mouthwashes, so keep that in mind when you're shopping for a good mouthwash for you or your children. Avoid antibacterial or antiseptic mouthwashes. Would you be willing to take an antibiotic twice a day, every day, for your body to prevent disease? Of course not. Why would you do that for your mouth?

Similarly, avoid acidic mouthwashes. You can and should determine the pH by using testing strips, which is quick and inexpensive, or use a more accurate pH meter. Look for an alkalizing mouthwash with a pH above 7.

Mouth Spray

Now that we've come this far together, I'm going to let you in on one of my favorite little secret weapons in combatting oral diseases. In fact, it's so little that you can literally fit it in your pocket, in your purse, or in your child's backpack or lunchbox, and carry it everywhere. Using the *right* mouth spray is one of my favorite ways to balance the pH of saliva throughout the day and thus keep the mouth healthy and happy.

Of course, I'm not talking about the terrible mouth sprays people typically use to mask their bad breath; if you use those, please see my comments in the mouthwash section.

You can think of a mouth spray as sort of a light mouthwash, with somewhat similar ingredients but formulated so you don't have to spit it out and can use it at any time. How does that little change make such a big difference?

Let's do a quick review of the demineralization and remineralization cycles. Every time you eat or drink, the pH of saliva drops and becomes acidic for about 30–60 minutes, and this is when the enamel is most vulnerable to damage. Since you shouldn't brush your teeth when they're in this weakened state, this is where an effective mouth spray can safely help balance the pH.

There is typically a 16-hour span between the morning oral care routine and the evening one, so you can use a pH balancing mouth spray after every meal to alkalize the saliva. And because it's really the only oral care product that you can take everywhere and use anywhere, you can help keep the mouth in peak health throughout the day. If your mouth spray can also provide prebiotics to promote the healthy microbes, even better still.

Toothbrush

A brush is a brush, right? Far from it!

Toothbrushes have come a long way since ancient civilizations used a chew stick or even since the invention of the modern toothbrush in 1938. Today, adult toothbrushes are beautifully designed with lots of choices for the bristles and high-tech electronics to help with every aspect of brushing. Children's toothbrushes are designed around their favorite superheroes, they play music, and some even use technologies like augmented reality to educate and make brushing fun.

Choosing the right toothbrush is important since it's a crucial tool in your oral care routine. Think of it this way: your child will spend over 2,000 hours brushing their teeth in their lifetime, so it's important to help them build positive associations with their oral care early on. That's why I recommend letting your child pick out a toothbrush that they love so they will enjoy and look forward to brushing.

Always use soft or ultra-soft bristles such as PBT (polybutylene terephthalate) or soft nylon. Abrasive bristles (such as hard nylon or charcoal brushes) can scratch the enamel, which makes it more susceptible to accumulating microbes. Damaged teeth are also more sensitive.

TIPS TO MAKE BRUSHING A SUCCESS:

- Change your brush every one to three months. Frayed and broken bristles are not as effective in cleaning your teeth. Softer bristles require changing more often since they lose their effectiveness sooner.

- Manual and electric toothbrushes are both fine, so choose whichever one you and your children prefer and are more likely to use consistently.
- Use a musical toothbrush to make it more enjoyable and to extend brushing times. Children brush 73 percent longer when brushing to music.
- Brush for two minutes first thing in the morning and two minutes in the evening before bedtime. This provides adequate time to brush all the teeth and for the toothpaste to do its magic.
- Don't press too hard when brushing.
- Keep the brush at a 45-degree angle to teeth and gums and brush in small circles. Brush all surfaces of every tooth.
- Help your children with their brushing. Five-year-olds brush only 25 percent of their teeth, and 11-year-olds only brush 50 percent of their teeth.
- Brush your tongue, too, or better yet, use a tongue scraper. Toothbrush bristles are not really designed to reach and eliminate the bacteria on the tongue that can cause bad breath and interfere with taste.
- Wait at least 30–60 minutes after meals before brushing.
- Store your brush properly. Avoid keeping them near other people's brushes. Avoid keeping them near toilets. Change them after a sickness.

Floss

Be honest: How often do you or your children floss? If you ask any dentist, they'll tell you their patients exaggerate (dare I say lie?) about flossing more than any other subject. Most people don't like flossing, and why should they? The typical over-the-counter floss makes it a terrible experience. You have to wrap it around your fingers, which can be uncomfortable. Then, just when you're ready to floss, it snaps in between your teeth and injures your gums, and you have a sink full of blood. Who needs that?

You certainly don't, but you do need a *good* floss.

Everyone must floss at least once a day because it's really the only tool that can get in between the teeth, where the harmful bacteria can hide. Children or parents who have difficulty using a regular floss can use a disposable flosser, with a handle, which makes it easier to get floss between the teeth. Children need help with flossing typically much longer than brushing, so you may need to floss for them for years before they're ready to do it on their own.

So now that you know you need to floss, let's learn how to choose a great one. As with everything else you put in your mouth and the mouths of your children, safety must be a priority. In 2019, a study claimed that some flosses expose people to toxic chemicals known as PFAS. Teflon is a brand name for nonstick cookware called polytetrafluoroethylene (PTFE), part of the family of PFAS chemicals.[57] Similarly, some flosses use waxes that are less than safe or use artificial flavors that we should all avoid.

A healthy mouth should never bleed while flossing. So, if you see blood in the sink, you either have gum disease or you're using a floss that's too harsh and irritating on the gums. Having said that, there are plenty of safe and effective floss materials such as silk, nylon, or polyester, and waxes like beeswax, that you and your children can comfortably enjoy while caring for your mouths.

CONNECTING THE DOTS

Below is a quick reference guide for Mouthrageous™ ingredients you should avoid at all costs, as well as a list of Mouthstanding™ ingredients that should be a part of your family's daily oral care routine.

Toothpaste, mouthwash, mouth spray	MOUTHRAGEOUS™	MOUTHSTANDING™
MICROBIOME	Antibacterial ingredients Triclosan Alcohol Essential oils	Prebiotics
SWEETENERS	Aspartame Saccharin Sorbitol	Xylitol Erythritol
FOAMING AGENTS	Sodium lauryl sulfate (SLS) Sodium laureth sulfate Propylene glycol Diethanolamine (DEA)	Quillaja saponaria extract (you can have a natural foaming factor like Quillaja to gently clean teeth, but you don't want so much foam that you can't see what you're doing)
FLAVORS & COLORS	Artificial flavors and dyes Titanium dioxide	Natural flavors and ingredients
pH BALANCE	Acidic (below 7 pH)	Alkaline (above 7 pH) pH balancing
Toothbrush	Abrasive bristles Poor quality	Soft or ultra-soft bristles Highest quality
Tongue scraper	Abrasive tongue brushes (avoid rough/stiff bristles) or scrapers (avoid uneven/sharp edges)	Gentle, yet effective tongue brushes (soft bristles) and/or scrapers (smooth/high quality)
Floss	Material: PTFE/PFOA/ PFAS Wax: Petroleum-based Artificial flavors	Material: Silk, nylon, or polyester Wax: Beeswax Natural flavors

IF YOUR MOUTH COULD TALK, IT WOULD SAY: "CARE FOR ME LIKE YOUR LIFE DEPENDS ON IT . . . BECAUSE IT DOES."

Think about your mouth for a moment. Obviously, it keeps you alive—you use it to eat, drink, and breathe. With your mouth you can kiss, taste, smile, sing an aria, start a riot, laugh when you're happy, cry when you're hurt, calm an infant, and shout from the rooftops. In very real ways, every day, your mouth makes life meaningful.

Your tongue, teeth, lips, and palate make it possible to form words, and with words you can communicate complex ideas. The human mouth is responsible for our civilization.

But, just as a mouth can make life meaningful, it can take life away. An unhealthy mouth is dangerous and goes hand in hand with an unhealthy body. It's both a cause and an effect.

Everything in the human life cycle is related to the health of the mouth: getting pregnant, having a healthy baby, sleeping soundly, doing well in school, achieving social success, finding a mate, getting a job, maintaining mental health, avoiding chronic or systemic disease, and aging well.

A healthy mouth is fundamental for a healthy life. In fact, it's hard to have a good life without a healthy mouth. Unfortunately, so many people don't understand how important it is, lack access to good care, or have personal obstacles that prevent them from caring for themselves (or all three). Poor oral health is a major crisis, and mounting evidence is showing us that it's one that goes by many other names—names like cancer, diabetes, bullying, ADHD, Alzheimer's disease, and sleep apnea.

It's time to recognize a root cause of all these systemic problems—to connect the dots so we can prevent disease before it starts, or at least treat it quickly and appropriately when it appears. It's time dentists and physicians started working together to treat patients. The mouth is the gateway to the body, and the body, in turn, affects the mouth. Dentists and physicians need to join forces and build bridges between the branches of health care.

THE GLORIOUS ECOSYSTEM THAT IS YOU

To paraphrase Albert Einstein, life is like riding a bicycle—to keep moving, you must keep your balance. Our bodies are all about keeping balance too. Remember, your body is a community. Microbes live on and inside of you all the time, and they keep you healthy or make you sick, depending on how well they are balanced and maintained.

The complex reality is that oral health depends on so many factors: the balance of the oral microbiome you start with, the oral microbiome of caregivers and family members, oral hygiene, diet, saliva, immune function, breastfeeding, airway health, mouth-breathing, oral care products, oral habits, environmental toxins, the growth and development of the structures of the mouth, dental and orthodontic factors—it's all a part of your complete oral health picture. Your microbiome interacts in a delicate and complex dance with your genetics (the DNA blueprint in your cells that you're born with) and epigenetics (the changes your environment makes to your DNA). All of this contributes to your oral health and whether or not you develop illnesses.

BEGIN WITH THE END IN MIND

Of course, none of us wake up one day suffering from obesity or cardiovascular disease. We don't suddenly end up with an anxiety disorder or chronic depression. The roots of most of life's challenges are in childhood, and if they're not addressed at the time, they gradually change the direction of our lives. These changes happen so slowly we might not notice, until one day, we don't recognize the person in the mirror.

What is it going to take to turn things around?

Culturally, it's going to take a massive push from all sides—from medical and dental professionals, and from our government. We will have to shift our thinking, values, and social norms to focus on whole body health. We must consider feeding bodies and microbes rather than taste buds. We must expect our government to support the wholesome food, access to health care, and clean environments we all need to thrive.

Amazingly, despite its dramatic impact on overall health, the topic of oral care has been mysteriously omitted from our cultural conversation around wellness and self-care. We have a million different types of content about weight loss. We have a massive fitness industry. We have several TV channels, dozens of cooking shows, and thousands of cookbooks about food. Where's the show about healthy teeth? Why no "mouth care" videos? When celebrities share their health secrets, brushing and flossing are never on the list. Maybe taking care of your teeth isn't sexy, but nice teeth sure are.

Oral care is critically important for living a healthy, successful life and for raising healthy, successful kids. Our government, culture, and the media are doing us all a disservice.

Changing the tide on public health is such a big challenge, it can feel overwhelming to even think about it. It's easy to stress oneself out fretting over the huge, seemingly intractable tragedy of public health in America. My advice is to start with what you have the power to change.

When you zoom in from the big picture to the family, and even further to the individual, change starts to feel easier. Just look at your own life. What habits are hurting you? Do you have personal obstacles like fear of dentistry or misplaced priorities that get in the way? How are you caring for your family's oral health at home? What are you eating that might be better left in the sealed, colorful bag on the grocery store shelf? Are you sleeping well?

Try this thought experiment: What if we could go back in time, knowing what we know now, and do things a little differently? What if we ate a little healthier, got a little more physical activity, reduced the toxicities around us, didn't overuse antibiotics, and took better care of our mouths . . . would things turn out differently?

Luckily, we have that opportunity. Although we cannot travel back through time (yet), we do have the ability to correct the health mistakes of our past. If you have children, you also have the power to create a very different story for them. And that story will continue because these habits are learned. Building this good health into the next generation is a revolutionary act! This is how cultures change.

CARE FOR YOUR CHILDREN'S MOUTHS
LIKE THEIR LIVES DEPEND ON IT

Before you became a parent, you were probably pretty focused on yourself. Maybe you worried about getting a good education, landing a solid career, or finding that special person. Then you had your child, and everything changed. All of your concerns converged around a single, shining feeling: "My child is everything." How do I know this about you? Because this is exactly what happened to me too.

This isn't to say being a parent is all joy. It's hard work: late-night feedings, changing endless diapers, finding the energy to stay engaged with your baby, striving to be a good role model, and doing everything you can to make the right decisions. As kids get older, you are responsible for finding them the right doctors, sending them off to school, figuring out how to advocate for them with teachers and other kids, and hoping they don't get bullied or suffer from low self-esteem. Being a parent is tough. But I hope the information in this book will make it just a little easier.

By now you know a big secret that most people don't: the mouth is the key to whole body health. It's the key to psychological health too. Everything starts here. This isn't taught in schools, though I hope someday it will be. But that doesn't make it any less true. This is powerful information that can quite literally change your family's life.

Not only is a child's mouth the gateway to their body, it's also malleable and flexible, just like the child. As such, childhood is the best time to establish lifelong habits, and to address many early issues that could cause big problems down the road. Raising children to have good oral care habits is the ultimate example of preventative medicine.

KIDS AND PARENTS, PARENTS AND KIDS

It's so common for me to see parents putting their kids first and totally ignoring their own oral health. The American Dental Association reports that parents consistently put their kids' dental care before their own. Often, parents feel like they can manage good care for only one family member, and that family member is going to be the child. I understand—being a parent is hard, resources are limited, and parents want the best for their kids. But I always try to remind the parents I see in my office: having you around and healthy is what's best for your kid.

Caring for your child means being healthy enough to care for your child.

Maybe you're ready to go to the dentist, but you're just not sure how to fit it in. And no wonder. Today's parents work more hours than ever. The average full-time American worker puts in 47 hours per week, with 39 percent working more than 50 hours. There are more single-parent households than ever. In 1960, single parents headed only 9 percent of families with children. By 2020, that number had increased to 25 percent.[1] Today, more mothers are working outside the home; in fact, 77 percent of women with children under age six are in the labor force. For women with children aged 6–17, participation increases to 81 percent.[2]

As we all know, raising children is an enormously time-consuming task. Many parents of very young children spend more time caring for them than they do in their full-time jobs, and even when their children reach school age, parents often spend more than 10 hours a week shuttling them to and from school and extracurricular activities. Work/life balance is imbalanced, and schedules are so hectic that time has become the most precious commodity of all.

Parents often feel that they have time either for their child's dental appointment or their own, but not both. In my practice, parents who bring their kids for dental checkups ask, "Can't you see me too?" After years of hearing that question on a daily basis, we launched Parent Dentistry.

Consider this: American kids miss 51 million hours of school each year due to dental issues. Imagine the financial strain this puts on parents. Each time a child has to stay home, a parent or caregiver must take off work, which can lead to reduced earnings or even loss of employment. Now, imagine if we could get those 51 million hours back!

THE TRUE COST OF DELAYED CARE

Many of my patients understand that oral health is critical, but they still feel like it's a pipe dream. They just don't think it's something they can afford. I know it can be burdensome to pay for oral care, even just preventative care. But it's so essential to avoid huge expenses down the line—including the expense of a life lost too soon. Preventative care is far cheaper than the alternative. And it's far safer.

If tooth decay is allowed to progress, it can become a life-threatening emergency—something families will *have* to address if they want their loved one to survive. At that point, treatment is likely to be orders of magnitude more expensive than preventative care would have been.

For example, once decay reaches the pulp of the tooth, it can cause infection requiring emergency surgery and hospitalization, sometimes amounting to tens of thousands of dollars. Infected teeth can spread infection to ears, sinuses, and organs such as the heart and kidneys. Often, antibiotics have no effect on these other infections until the dental decay is treated. I don't say this to fear-monger—it's just the truth! Problems in the mouth can be life threatening, and not just because they contribute to long-term chronic health problems. They can progress quickly, causing near-term problems too. Sometimes, dire ones.

A well-publicized case illustrates this danger. A 12-year-old boy in Maryland, healthy in every other way, had a cavity. It went untreated, and an abscess developed. By the time he finally made it to the doctor, the bacteria from the abscess had spread to his brain. He endured two surgeries and weeks of hospital care totaling about $250,000 in an attempt to save his life, but it was too late. The boy died—a tragedy that could have been prevented with an $80 tooth extraction.[3]

In 2017, a simple untreated tooth infection spread to a man's lungs and killed him. He was a 26-year-old father living in California.[4]

These may be extreme examples, but even in my own practice, we routinely see patients with severe dental infections that require immediate treatment or hospitalization. Happily, after proper dental care, most of these patients return to good health and make a full recovery, but at what cost?

Still, even when they understand the dangers, many parents just can't afford dental care. For those parents, at our practice, we offer as much financial help as we can. By seeing Medicaid patients and helping others as much as possible, we hope to cover the gaps. But it's clear that more work needs to be done in this country to support access to dental care for all families.

Trouble affording care rears its ugly head in so many unfortunate ways. For example, when it comes to orthodontics, many families delay and avoid treatment because of the cost, only to pay so much more down the road to correct a problem that is now far harder to treat. Sometimes parents opt for the cheapest possible treatment, not realizing there are many hidden costs, or that the treatment is far less likely to be effective. Often, parents simply wait too long to see the orthodontist. If they had sought treatment earlier, when the child was younger and the bones of the jaw had yet to fuse or stop growing, they could have addressed the problem with easier, cheaper interventions.

The same principles apply when it comes to your daily oral care routine at home. If you use the right tools and do it routinely, and well, you'll prevent many problems before they even begin—from cavities to gum disease to chronic, systemic illness. I often marvel at how many health care dollars are saved in a lifetime when someone brushes, flosses, uses safe and effective oral care products, and gets regular cleanings. Not to mention the extra years that person gets to enjoy.

Always, this story returns to the same place: prevention. It's the safest, easiest, cheapest, and best road to oral care and general health. Hands down.

WHY DOES GOING TO THE DENTIST FEEL SO *HARD*?

Even people who have adequate resources, are well educated, and deep-down care very much about the health and appearance of their teeth, don't go to the dentist. Maybe it's because these are people who lacked good care as children and that was what they were taught. Maybe they've decided their mouths are lost causes. Or maybe they're just too afraid of what that dentist might find: they are embarrassed, afraid to be shamed about bad habits, or afraid of the expensive procedures the dentist will recommend. In many cases, they're simply afraid of the dentist.

A friend of mine has a very hard time going to the dentist. He never had a particularly bad experience, and his mouth isn't in terrible shape. But he feels guilty about never flossing, stressed about the cost, and afraid of what he might find out about his oral health. Chances are, he wouldn't find out anything terrible. He's never had cavities before. His teeth *look* fine—they seem strong and aren't discolored or sore. His gums aren't swollen or red. Yet he has a mental block about the dentist and just can't seem to get himself to go! He's relatively healthy now—it's prime time for the preventative measures that will keep him healthy for many years to come. If he can't get past this, he might not stay that way.

Going to the dentist is up there with death and public speaking as the most feared thing in life, often born in bad childhood experiences. This cycle repeats as parents pass on that fear to their kids through epigenetics or psychological reinforcement. And sometimes, as in the case of my friend, it's an irrational fear based on nothing but a general apprehension.

If you *did* have a bad experience at the dentist when you were a child, it's probably still with you. About 30 to 40 million Americans suffer from dental phobias.[5] Thirty-six percent of American adults avoid dental appointments because of these fears. Often, it's a parent's worry that triggers fear in their children. Either that parent expresses apprehension openly, or they instill it in quieter ways—by avoiding dental appointments, or by bringing stress and anxiety to the appointments, making them unpleasant for the child. Other times, a child has a bad experience, and that creates a negative association for life.

However, dentistry has changed dramatically. It doesn't have to be a negative experience anymore. Dentists have access to advanced tools and modern technology that make diagnosis and treatment a lot easier and quicker. We use calming techniques and entertainment such as movies, toys, and games to make the experience pleasant, and even fun for patients.

This is a fundamental part of the care we provide and it's a building block of my practice, The Super Dentists. I believe kids and their parents can enjoy going to the dentist, and love caring for their teeth. It's all about the experience.

This is a big reason why it's so important to take your children to a pediatric dentist. They are specialists—both technically great at working on kids and sensitive to the emotional experiences of children. Pediatric dentists know the power childhood experiences have in shaping a lifetime of health (or a lack thereof). Many of us dedicate ourselves to creating fun, magical dental experiences for our patients.

Our aim is to break the negative feedback loop. Because the longer someone waits, the harder it is to take action. Negative consequences have a tendency to snowball. Anxiety about the dentist leads to missed appointments and yellow, unattractive teeth. Shame about that smile leads to low self-esteem, which causes trouble at work, resulting in less access to insurance. Even more time goes by without treatment, and before you know it, minor problems have become major ones.

IF YOU THINK YOU CAN'T, YOU MUST

Sometimes, the only way to break the loop is to act.

I have an example of this from my own life. I've always been interested in the motivational self-help genre, and a few years ago, I decided to join a workshop with a well-known leader in the field. One day during the workshop, we had to do a physical exercise that, frankly, terrified me. It involved climbing up a 40-foot pole, at the top of which was a small platform. You were supposed to climb from the top of the pole to the platform without anything to hold on to, 40 feet up in the air.

I made it up the pole just fine, but when I got to the top, I froze. My legs were shaking. I broke out in a cold sweat. There was no way I could step from the top rung of the pole, over the edge of the platform, without anything to hold onto! No. Way. I clung to the pole and just kept thinking "I can't! I can't!" over and over. Then, somehow, something shifted. I realized this was the whole point. When you think you can't, you must. You have to act. You're out on a limb (or up a pole, in my case), and acting is the only way forward. As my legs shook uncontrollably, I took that final step. I scrambled onto that platform. Once I got up there, the nerves vanished. I stopped shaking. Everything felt easier.

If you think you can't, you must. Schedule an appointment. Take that final step. Act. I bet everything will feel easier once you do. The only way out is straight through.

PARENTAL PRIORITIES

Clearly, a parent's deepest desire is to help their child grow into a confident, healthy young adult. But even parents who have the means, and who generally do a good job of providing health care to their kids, and who don't have a particular fear of the dentist, still don't make dental care a priority. It's a clear result of the terrible job we've done educating the public about the importance of good oral care.

It still shocks me to see parents with resources balk at the cost of braces. "Well, that's a lot of money," they say. "I'll think about it and get back to you."

Since we're talking dollars and cents, let's consider oral health in terms of a financial investment. If I offered you the chance to invest $7,000 with a return of $250,000, you would probably leap at the opportunity. Well, $7,000 is the average cost of braces, and $250,000 is how much more a person with a healthy mouth and a nice smile is likely to earn in a lifetime (for those of you looking for a calculator, that's more than 140 percent *annual* return over a 25-year career).[6]

Studies also show that for every $1 invested in preventive dental care, a person should expect to save from $8 to $50 in future

restorative and emergency treatments (not to
mention the cost savings that you and your
child could enjoy over a lifetime free from
chronic disease).[7] When you combine this
with the dramatic impact a healthy mouth
and beautiful smile has on a person long
term, timely care is clearly the best choice.

INFORMATION OVERLOAD

We've got a child and parent health crisis, no
time, not enough money, and tons of anxiety to contend with. How
about an endless stream of bad information too? The explosion of
information on the internet in recent years has left everyone wondering whom to trust. Sometimes we have no way of knowing if
we're hearing the opinion of a highly trained doctor connected to a
world-renowned research facility or some opinionated blogger in
their pajamas with no medical or dental training.

To make things more difficult, even when information comes
from genuine experts, it's often contradictory. As it relates to oral
health, there is actually one good reason for the dissonance: dentistry is partly art and partly science, so doctors can have honest
disagreements.

But all this disagreement is also because well-meaning
researchers may have studied and evaluated one particular issue
or effect, but they don't consider the full picture we're discussing
here: the physical, psychological, social, and economic implications of a treatment over a lifetime.

It's no wonder that after being barraged by so much contradictory information, people often feel like saying, "Forget it. I can't
deal with it!"

My approach connects the dots. It is my unwavering conviction
that everything in the body is related to everything else.

EASY OR SIMPLE ISN'T ALWAYS BEST

Easy is great when it comes to some things—like preventative care. But sometimes the easy treatment just isn't going to be effective, and with your health it's important to evaluate *effectiveness* first, ease or cost second. This is never truer than with the infamous "wait and see" approach. For example, it may be easier (in the moment) to wait and watch than to treat a tongue-tie. But when that baby's speech is delayed, or the mouth doesn't grow and develop correctly, real damage is done.

As a parent, getting the right care, and timely care, can be one of the most important decisions you make for your child's future. When it comes to children, there is always a window of opportunity during which treatment will get ideal results, and once that window closes, treatment options may become limited or more invasive, and the result can be significantly compromised.

Not long ago, I treated a teenager who had suffered through years of braces and multiple surgeries on both of her jaws because her mother was told to wait and watch some baby teeth that were ankylosed (fused to the bone) when she was younger. When she finally came to see me as a teen, her fused baby teeth had impeded the growth of her jaws and resulted in a significant malformation of her face. This is something that could easily have been prevented by treating a few baby teeth and engaging in a short (and significantly less costly and less invasive) orthodontic treatment when the issue was first discovered.

Another often unexpected (but common) example of how waiting can impact a child relates to overbites. A child with an "overjet" (where the upper jaw is too far forward—often called an overbite) can break their front teeth remarkably easily. They stick out and don't have the support of the bottom teeth behind them, so any rough-and-tumble impact can crack them right in half (it's like having long nails—they're inherently more prone to breakage).

Fixing the overjet later in life is not only more difficult, sometimes it can even be impossible without surgery, and the person will end up with restorations on the front teeth that may never look truly natural.

Take your kids to the pediatric dentist before they turn one (ideally shortly after birth, before the first baby teeth erupt), and keep taking them regularly. When the dentist suggests an early intervention, consider not just the immediate cost or difficulty, but the long-term costs and difficulties if the problem is allowed to progress. Your dentist or orthodontist should be able to explain all of this to you clearly, if you ask.

ENGAGE YOUR DENTAL CARE TEAM EARLY

By now you know your dentist is there for more than just filling cavities. Your orthodontist can do more than just straighten teeth. These are experts on oral care in all its various forms. They can help your children break bad habits that can negatively impact the growth of their mouths (like thumb-sucking or pacifier use) and help you establish better habits for your own oral care, with tips and tricks for making brushing and flossing a regular part of your routine. They can be your health partners, from conception and pregnancy, to helping ensure the healthy growth and development of your baby, to helping you and your children breathe and sleep better, and to maintaining the oral health necessary for you and your children to have longer, happier lives.

After having read this book, you know the importance of early intervention. And that what we've been told for years—kids shouldn't go to the orthodontist until they're 12 or 13—is bogus. The American Association of Orthodontists recommends a checkup with an orthodontic specialist no later than age seven. That's because by age seven, there are usually enough permanent teeth to evaluate the bite and dental space, as well as enough growth left to fix most bite issues. The maxillary sutures have also not started to fuse yet (they start fusing at around age eight). But that's only if there are no signs or symptoms. If there are other issues, then you should seek treatment much earlier.

I always get very sad when I see patients who come to me at 12 or older with crooked teeth, mouth-breathing, sleep issues, poor school performance, and behavioral challenges. Many have been told they should just wait for all permanent teeth to come in before getting braces.

Dentists and orthodontists can start helping your child's mouth grow correctly from a very young age. It may begin with a pediatric dentist revising a tongue-tie in an infant. Perhaps a few years later, an orthodontist puts in a palatal expander so that your child's palate grows larger, making room for teeth, aligning the jaws for a better bite, and opening the airway to prevent sleep apnea and other breathing problems. When you visit regularly, these health partners can step in as problems arise, preventing things like pulled teeth or even braces down the road, and helping to create a more attractive smile and face. As you know (and have heard me say before), an attractive smile goes hand in hand with good self-esteem, self-confidence, and overall success in life.

It's sort of ironic that the very things we associate with dentists and orthodontists—filling cavities and straightening teeth—are the things you may never have to deal with, provided you get your child in early and practice good preventative care. Maybe someday we will associate these professionals with overall healthy growth and development instead. That is certainly my hope!

If your family does not have access to a group dental practice where pediatric dentists, orthodontists, general dentists, and other specialists work side by side and monitor the growth and development of your children right from the start, then it's useful to remember the standard guidelines for timing for dental care in kids. It's my "1, 4, 7 Rule."

1—Take your child to a pediatric dentist no later than age one.

4—Work with your child's dentist to end any poor oral habits (like thumb-sucking) no later than age four.

7—Bring your child for a checkup with an orthodontist no later than age seven.

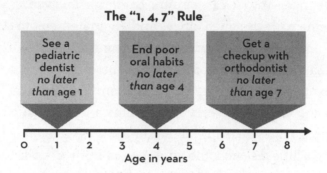

Again, these are outer limits of when children should be seen, and you should always seek advice or treatment if any issues occur earlier.

This timing is so important because it will allow you to address any emergent problems with the most efficient, effective, and least expensive interventions.

TECHNOLOGIES MATTER

There is some exciting cutting-edge tech in the dental world, and it dramatically improves outcomes, reduces pain, and makes the whole experience faster and easier. Just like in any profession, not all dental offices are created equal.

New technologies have changed everything you hate about going to the dentist. We now have low- to no-radiation cavity detection technologies, computerized anesthesia for increased comfort, digital oral scanners to eliminate gagging and improve accuracy, lasers to make procedures safer and faster with quicker healing, self-adjusting braces for fewer appointments and less pain, accelerated orthodontic treatment technologies for shorter treatments and better results, and so much more.

We have also made major strides in the experience of going to the dentist—now it can be fun, comfortable, even enjoyable! From theme park–designed offices and entertainment, to childcare services and cafés, dental offices have come a long way from the sterile, scary places many of us fear. Some of my patients' favorite perks at The Super Dentists include aromatherapy, massages, flavored and scented gloves, virtual-reality and augmented-reality experiences, and other *super* amenities.

Doing a little research into your dentist's office—to learn about the technological state of their office and equipment—can tell you a lot about the type of experience you're likely to have. You can save yourself and your family some pain (literally) by seeking out an office that values state-of-the-art equipment and techniques.

THE LIFE YOU GIVE YOUR KIDS

As a parent, good oral care habits are an invaluable gift you can give your kids. Parents who regularly take their children to the dentist and make it a normal part of growing up tend to instill that normalcy in their kids. If they help their kids practice good preventative care at home and provide them effective and playful tools, their kids enjoy taking care of their mouths and are less likely to fear dental procedures, because they have fewer of them. As adults, they are more likely to brush and floss and go to the dentist, because they always have. Oral health becomes a familial value and a priority, and that culture of care can continue on and on for generations of healthy children and adults.

It all starts with you.

There is so much hope in good care, but we can't ignore or forget the dark side here too: the silent, pervasive, damaging oral health situation afflicting so many kids and adults. I just can't sugarcoat it (nor would I . . . because I'm a dentist . . . see what I did there?).

THE LIFE YOU GIVE YOURSELF

What are you willing to do to get a 10-year extension on your life? What about another 15 years? That is so much more time to breathe, travel, find joy, master a craft, and watch your children and grandchildren grow up. What is that worth to you?

Remember, knowledge is power; but *applied* knowledge can be life changing. You've taken a big step toward empowering yourself and your family by reading this book. The next step—and the most important one—is applying what you have learned.

Oral health starts early, so teach your kids what you've learned here and stack the odds in their favor. But don't forget about yourself. Take care of your mouth. Lead by example. Help create positive associations with the dentist for your child while you heal from negative dental experiences in your past. Brush, floss, use appropriate oral care products, visit the dentist regularly, apply your new knowledge, and live happier, healthier, and longer. Enjoy your kids. Enjoy your life.

CALLING ALL PARENTS!

By reading this book, you have taken the first step toward improving the life of your family. These pages contain the key to your children's health and success.

Now it is time to take the next step and apply your knowledge. You can find cutting-edge innovation and all the right tools at

DRKAMIHOSS.COM

As a thank-you for taking this journey with us, readers of this book will receive a special gift to launch you and your kids on a path to health and prosperity.

A NOTE TO MY COLLEAGUES

This seems like the right moment to appeal to readers who also happen to be dental or medical professionals. Just as it isn't easy for our patients to seek care, it isn't easy for us to change the way we deliver it. But we need to broaden our knowledge and awareness.

For dentists and orthodontists, it's time we stopped just fixing symptoms by drilling, filling, pulling, and straightening teeth and started looking deeper. Our cornerstones should be prevention, education, collaboration, and comprehensive care that looks for root problems. We need to see signs of systemic disease when they appear, and work with physicians to address those problems together.

For physicians, it's time to stop feeling like the mouth is outside your jurisdiction and start feeling more empowered to recognize when it's in need of specialized care. The more we can cross-pollinate between our disciplines, the more our patients will benefit. It's not easy, but I truly believe it is the only way forward, given what we now know about the root causes of systemic diseases and how inextricably linked the mouth is with the rest of the body.

Colleagues, we need to look beyond our own specialties, to treat our patients as whole people. We also need to make going to the doctor or dentist fun and exciting, so that patients (both young and old) look forward to their visits and are emotionally engaged in taking good care of their mouths and bodies. I am ready and excited to be part of this new era in cooperative care. I welcome you to join me! I hope this book can help pave the way.

ACKNOWLEDGMENTS

Alone we can do so little; together we can do so much.
—Helen Keller

To my brilliant wife and partner, Dr. Nazli Keri. You shared innovative ideas, raised important issues, and helped steer the conversation to ensure the book delivers a compelling and convincing account of why everyone needs to take care of their mouth. This book wouldn't be possible without you, and I am eternally grateful for your insight and support.

To the caring doctors and crew members at The Super Dentists. Without your dedication, commitment to excellence, and unwavering belief in the mission of delivering high-quality oral care to all families, The Super Dentists would not exist.

To my wonderful patients. Over the course of more than two decades, The Super Dentists has had the privilege of earning the trust of hundreds of thousands of families. We are committed to transforming the dental experience and delivering the highest level of dental and orthodontic care to children, teenagers, and adults. Thank you for putting your lives in our hands.

To the extraordinary team at SuperMouth. Your uncompromising commitment to innovation, your refusal to accept the status quo, and your incredible work ethic have yielded a revolutionary line of oral care products that kids love using. I am so proud of our achievements.

To my colleagues in the health care profession. Your dedication to building a better and stronger culture of integrated oral and physical health care in America benefits us all.

To the talented and creative team behind this book. My deepest gratitude to Anni Murray, Alicia Dunams, and Beata Santora for guiding me through this journey. Your insight, wisdom, and attention to detail made this book possible. It takes a village and I'm so proud to be a part of this one.

To my literary agent and dedicated team at BenBella Books. Your vision for the book and faith in me was an inspiration throughout this process. I'm grateful for your expertise and determination.

NOTES

INTRODUCTION

1   Roizen, Michael. *RealAge: Are You as Young as You Can Be?* New York: William Morrow (1999).

2   Sleep Group Solutions. https://join.sleepgroupsolutions.com/.

3   Sujata, Tungare, and Arati G, Paranjpe. "Baby Bottle Syndrome." *StatPearls* (January 2021). https://pubmed.ncbi.nlm.nih.gov/30570970/.

4   Centers for Disease Control and Prevention. "Chronic Diseases in America" (2021). https://www.cdc.gov/chronicdisease/resources/infographic/chronic-diseases.htm.

5   Alzheimer's Association. "Facts and Figures" (2021). https://www.alz.org/alzheimers-dementia/facts-figures?.

6   "History of Dentistry." American Dental Association (2020). https://www.ada.org/en/member-center/ada-library/dental-history.

7   Ibid.

8   Otto, Mary. *Teeth: The Story of Beauty, Inequality, and the Struggle for Oral Health in America.* New York: The New Press (2017).

9   Logan, Henrietta L., Yi Guo, Virginia J. Dodd, Christine E. Seleski, Frank Catalanotto. "Demographic and Practice Characteristics of Medicaid-Participating Dentists." *U.S. National Library of Medicine, National Institutes of Health* (June 16, 2015). https://www.ncbi.nlm.nih.gov/pmc/articles/PMC4469354/.

10 American Dental Association. "Oral Health and Well-Being in the Unites States" (2021). https://www.ada.org/~/media/ADA/Science%20and%20Research/HPI -/OralHealthWell-Being-StateFacts/US-Oral-Health-Well-Being.pdf.

CHAPTER 1

1   United Health Foundation. "America's Health Rankings, 2020 Annual Report" (2020). https://www.americashealthrankings.org/learn/reports/2020-annual-report/international-comparison.

2   Ibid.

3   Centers for Disease Control and Prevention. "Oral Health" (2021). https://www.cdc
     .gov/oralhealth/conditions/index.html.

4   Glick, Michael. *The Oral-Systemic Health Connection*. Batavia, IL: Quintessence
     Publishing Company (2019), 44.

5   Curatola, Gerald P., and Diane Reverand. *The Mouth-Body Connection*. New York:
     Center Street (2017), 13.

6   Lin, Steven. *The Dental Diet*. Carlsbad, CA: Hay House Publishing (2019).

7   Ibid.

8   Ibid.

9   Kutsch, Kim, and Robert J. Bowers. *Balance: A Guide for Managing Dental Caries
     for Patients and Practitioners*. San Diego: Illumina Press (2012).

10   New Hampshire Department of Health and Human Services. "How Much Sugar
     Do You Eat? You May Be Surprised!" (2020). https://www.dhhs.nh.gov/dphs/nhp
     /documents/sugar.pdf.

11   Lin, Steven. *The Dental Diet*.

12   Kutsch, Kim. *Why Me? The Unfair Reason You Get Cavities and What to Do About It*.
     WellPut Custom Content (January 2020).

13   Benito, Eva, Cemil Kerimoglu, Binu Ramachandran, Dieter Edbauer, Camin Dean,
     and Andre Fischer. "RNA-Dependent Intergenerational Inheritance of Enhanced
     Synaptic Plasticity After Environmental Enrichment." *Cell Reports* (March 2019),
     546–54. https://www.cell.com/cell-reports/fulltext/S2211-1247(18)30404-2.

14   Reynolds, Gretchen. "Do Fathers Who Exercise Have Smarter Babies?" *New York
     Times* (May 9, 2018). https://www.nytimes.com/2018/05/09/well/move/do-fathers
     -who-exercise-have-smarter-babies.html.

15   Dias, Brian G., and Kerry J. Ressler. "Parental Olfactory Experience Influences
     Behavior and Neural Structure in Subsequent Generations." *Nature Neuroscience*
     (December 1, 2013), 89–96. https://www.nature.com/articles/nn.3594.

16   Whittaker Brown, Stacey-Ann, Bian Liu, and Emanuela Taioli. "The Relationship
     Between Tobacco Smoke Exposure and Airflow Obstruction in US Children."
     Chest, 2018; 153 (3): 630 DOI: 10.1016/j.chest.2017.10.003.

17   Kaati, G., L. O. Bygren, and S. Edvinsson. "Cardiovascular and diabetes mortality
     determined by nutrition during parents' and grandparents' slow growth period."
     Eur J Hum Genet. 2002 Nov; 10 (11): 682–8. https://pubmed.ncbi.nlm.nih.gov
     /12404098/.

18   Mahmoud, Abeer M., and Mohamed M. Ali. "Methyl Donor Micronutrients That
     Modify DNA Methylation and Cancer Outcome." *Nutrients* (March 2019). https://
     www.ncbi.nlm.nih.gov/pmc/articles/PMC6471069/.

19   Onusic, Sylvia. "Nutrigenomics." *Price-Pottenger* (July 19, 2016). https://price
     -pottenger.org/journal_article/nutrigenomics/.

20   Lieberman, Daniel E. *The Evolution of the Human Head*. Cambridge, MA: Belknap
     Press (2011).

21   Lin, Steven. *The Dental Diet.*

22   Lieberman, Daniel E. *The Evolution of the Human Head.*

23   Ibid.

24   Ibid., 224.

25   Ibid.

26   Clark, Laura. "Before Agriculture, Human Jaws Were a Perfect Fit for Human Teeth." *Smithsonian Magazine* (February 6, 2015). https://www.smithsonianmag .com/smart-news/how-dawn-farming-changed-our-mouths-worst-180954167/.

27   Crash Course World History. "The Agricultural Revolution: Crash Course World History #1." *YouTube* (January 26, 2012), 11:10. https://www.youtube.com/watch ?v=Yocja_N5s1I.

28   Ryan, Timothy M, and Colin N. Shaw. "Gracility of the Modern Homo Sapiens Skeleton Is the Result of Decreased Biomechanical Loading." *Proceedings of the National Academy of Sciences* (December 22, 2014). https://www.sciencedaily.com /releases/2014/12/141222165033.htm.

29   "Evolution of Corn." *Genetic Science Learning Center, University of Utah.* https:// learn.genetics.utah.edu/content/selection/corn/.

30   Plumer, Brad. "Here's What 9,000 Years of Breeding Has Done to Corn, Peaches, and Other Crops." *Vox* (May 12, 2016). https://www.vox.com/2014/10/15/6982053 /selective-breeding-farming-evolution-corn-watermelon-peaches.

31   Ibid.

32   Boyd, Kevin L. "Darwinian Dentistry: An Evolutionary Perspective." *Price-Pottenger Journal of Health and Healing* (Winter 2011). https://price-pottenger.org/journal _article/darwinian-dentistry-an-evolutionary-perspective/.

33   Ruggles, Steven. "Patriarchy, Power, and Pay: The Transformation of American Families, 1800–2015." *Demography* (December 2015), 1797–1823. https://www.ncbi .nlm.nih.gov/pmc/articles/PMC5068828/.

34   Price, Weston A. *Nutrition and Physical Degeneration.* Long Grove, CA: Price-Pottenger Nutrition Foundation (2009).

35   Larson, Rebecca. "Queen Elizabeth's Rotten Teeth." *Tudors Dynasty* (December 2017). http://www.tudorsdynasty.com/queen-elizabeths-rotten-teeth/.

36   Zimmer, Carl. "The Evolution of Cavities." *National Geographic* (December 21, 2012). https://www.nationalgeographic.com/science/phenomena/2012/12/21/the -evolution-of-cavities/.

37   Stothart, Mason R., Colleen B. Bobbie, Albrecht I Schulte-Hostedde, Rudy Boonstra, Rupert Palme, Nadia C.S. Mykytczuk, and Amy E.M. Newman. "Stress and the Microbiome: Linking Glucocorticoids to Bacterial Community Dynamics in Wild Red Squirrels." *Biology Letters, U.S. National Library of Medicine, National Institutes of Health* (January 2016). https://www.ncbi.nlm.nih.gov/pubmed /26740566.

38   Matsumoto, Megumi, Ryo Inoue, Takamitsu Tsukahara, Kazunari Ushida, Hideyuki Chiji, Noritaka Matsubara, and Hiroshi Hara. "Voluntary Running

Exercise Alters Microbiota Composition and Increases n-Butyrate Concentration in the Rat Cecum." *Bioscience, Biotechnology, and Biochemistry, U.S. National Library of Medicine, National Institutes of Health* (February 2008), 572–76. https://www.ncbi.nlm.nih.gov/pubmed/18256465.

39   Savin, Ziv, Shaye Kivity, Hagith Yonath, and Shoenfeld Yehuda. "Smoking and the Intestinal Microbiome." *Archives of Microbiology* (July 2018), 677–84. https://www.ncbi.nlm.nih.gov/pubmed/29626219.

40   Tun, Hein M., Theodore Konya, Tim K. Takaro, Jeffrey R. Brook, Radha Chari, Catherine J. Field, David S. Guttman, Allan B. Becker, Piush J. Mandhane, Stuart E. Turvey, Padmaja Subbarao, Malcolm R. Sears, James A. Scott, Anita L. Kozyrskyj, the CHILD Study Investigators. "Exposure to Household Furry Pets Influences the Gut Microbiota of Infants at 3–4 Months Following Various Birth Scenarios." *Microbiome* (April 6, 2017). https://www.ncbi.nlm.nih.gov/pmc/articles/PMC5382463/.

CHAPTER 2

1   Dasanayake, A.P., S. Gennaro, K. D. Hendricks-Muñoz, and N. Chhun. "Maternal Periodontal Disease, Pregnancy, and Neonatal Outcomes." *American Journal of Maternal/Child Nursing, US National Library of Medicine, National Institutes of Health* (January–February 2008), 45–49. https://www.ncbi.nlm.nih.gov/pubmed/18158527.

2   Paju, Susanna, Juha Oittinen, Henna Haapala, Sirkka Asikainen, Jorma Paavonen, and Pirkko J. Pussinen. "*Porphyromonas Gingivalis* May Interfere with Conception in Women." *Journal of Oral Microbiology, US National Library of Medicine, National Institutes of Health* (June 12, 2017). https://www.ncbi.nlm.nih.gov/pmc/articles/PMC5508391/.

3   Kellesarian, Sergio Varela, Michael Yunker, Hans Malmstrom, Khalid Almas, Georgios E. Romanos, and Fawad Javed. "Male Infertility and Dental Health Status: A Systematic Review." *American Journal of Men's Health* (November 2018), 1976–84. https://www.ncbi.nlm.nih.gov/pmc/articles/PMC6199424/.

4   Uwambaye, Peace, Cyprien Munyanshongore, Stephen Rulisa, Harla Shiau, Assuman Huhu, and Michael S. Kerr. "Assessing the Association Between Periodontitis and Premature Birth: A Case-Control Study." *BMC Pregnancy and Childbirth* (March 12, 2021). https://bmcpregnancychildbirth.biomedcentral.com/articles/10.1186/s12884-021-03700-0.

5   Glick, Michael. *The Oral-Systemic Health Connection*. Batavia, IL: Quintessence Publishing (2014), 206.

6   Gelb, Michael, and Howard Hindin. *Gasp: Airway Health—The Hidden Path to Wellness*. Scotts Valley, CA: CreateSpace Independent Publishing Platform (2016), 115.

7   Franklin, K. A., P. A. Holmgen, F. Jonsson, N. Poromaa, H. Stenlund, and E. Svanborg. "Snoring, Pregnancy-Induced Hypertension, and Growth Retardation

of the Fetus." *Chest, US National Library of Medicine, National Institutes of Health* (January 2000). 137–41. https://www.ncbi.nlm.nih.gov/pubmed/10631211.

8 Gelb, Michael, and Howard Hindin. *Gasp: Airway Health—The Hidden Path to Wellness.*

9 Skouteris, Christos A. (Editor). *Dental Management of the Pregnant Patient.* Hoboken, NJ: Wiley-Blackwell (2018).

10 Ibid.

11 Marchi, Kristen S., Susan A. Fisher-Owens, Jane A. Weintraub, Zhiwei Yu, and Paula A. Braverman. "Most Pregnant Women in California Do Not Receive Dental Care: Findings from a Population-Based Study." *Public Health Reports, U.S. National Library of Medicine, National Institutes of Health* (November–December 2010), 831–42. https://www.ncbi.nlm.nih.gov/pmc/articles/PMC2966664/.

12 Glick, Michael. *The Oral-Systemic Health Connection.*

13 Ibid., 201, ref. 7.

14 American Academy of Periodontology. "Expectant Mothers' Periodontal Health Vital to Health of Her Baby." (2021). https://www.perio.org/consumer/AAP_EFP_Pregnancy.

15 Seow, W.K. "Effects of Preterm Birth on Oral Growth and Development." *Australian Dental Journal* (April 1997), 85–91. https://pubmed.ncbi.nlm.nih.gov/9153835/.

16 Curatola, Gerald P., and Diane Reverand. *The Mouth-Body Connection: The 28-Day Program to Create a Healthy Mouth, Reduce Inflammation and Prevent Disease Throughout the Body.* New York: Center Street (2017).

17 Mahesh, Mahadevappa. "Computed Tomography Dose (CT Dose)." *Radiological Society of North America* (2021). https://www.radiologyinfo.org/en/info/safety-xray.

18 Yong, Ed. "Breast-Feeding the Microbiome." *New Yorker* (July 22, 2016). https://www.newyorker.com/tech/annals-of-technology/breast-feeding-the-microbiome.

19 Lin, Steven. *The Dental Diet.* Carlsbad, CA: Hay House Publishing (2019), ref. 5.

20 Moral, Angel, Ignasi Bolibar, Gloria Seguranyes, Josep M. Ustrell, Gloria Sebastiá, Cristina Martínez-Barba, and Jose Ríos. "Mechanics of Sucking: Comparison Between Bottle Feeding and Breastfeeding." *BMC Pediatrics, U.S. National Library of Medicine, National Institutes of Health* (February 2011). https://www.ncbi.nlm.nih.gov/pmc/articles/PMC2837866/.

21 "Breastfeeding and the Use of Human Milk." *Pediatrics* (March 2012), 827–41. https://pediatrics.aappublications.org/content/129/3/e827.

22 "AAP Reaffirms Breastfeeding Guidelines." *American Academy of Pediatrics.*

23 Chua, S., S. Arulkumaran, I. Lim, N. Selamat, and S. S. Ratnam. "Influence of Breastfeeding and Nipple Stimulation on Postpartum Uterine Activity." *British Journal of Obstetrics & Gynaecology* (September 1994) 804–5. https://obgyn.onlinelibrary.wiley.com/doi/abs/10.1111/j.1471-0528.1994.tb11950.x.

24 Skouteris, Christos A. (Editor). *Dental Management of the Pregnant Patient.*

25 Ibid.

26   Dewey, K. G., M. J. Heinig, and L. A. Nommsen. "Maternal Weight-Loss Patterns During Prolonged Lactation." *American Journal of Clinical Nutrition* (August 1993), 162–66. https://www.ncbi.nlm.nih.gov/pubmed/8338042.

27   Gaskin, Ina May. *Ina May's Guide to Breastfeeding: From the Nation's Leading Midwife,*: Bantam Books (2009).

28   Hill, Rebecca R., Christopher S. Lee, Britt F. Pados. "The Prevalence of Ankyloglossia in Children Aged <1 year: A Systematic Review and Meta-analysis." *Pediatric Research*. (November 2020). https://pubmed.ncbi.nlm.nih.gov/33188284/.

29   Kotlow, Lawrence. *SOS 4 TOTS: Tethered Oral Tissues, Tongue-Ties & Lip-Ties.* Troy, NY: The Troy Book Makers (2016).

30   Garcia, Sandra E. "Texas Boy Speaks Clearly for First Time After Dentist Discovered He Was Tongue-Tied." *New York Times* (September 29, 2018). https://www.nytimes.com/2018/09/29/us/tongue-tied-boy-speaks.html.

31   Nolan, Caitlin. "Texas Boy Thought to Be Nonverbal Can Speak After Dentist Discovers He's 'Tongue-Tied.'" *Inside Edition* (September 25, 2018). https://www.insideedition.com/texas-boy-thought-be-nonverbal-can-speak-after-dentist-discovers-hes-tongue-tied-47101.

32   Haller, Leslie A., and Theodore Brown. "Upper-Lip Frenum as a Predictive Marker for Unexpected and Unexplained Asphyxia in Infants." *Journal of Rare Disorders: Diagnosis & Therapy* (March 10, 2016). http://raredisorders.imedpub.com/upperlip-frenum-as-a-predictive-marker-for-unexpected-and-unexplained-asphyxia-in-infants.php?aid=8758.

33   Guilleminault, Christian, and Yu-Shu Huang. "From Oral Facial Dysfunction to Dysmorphism and the Onset of Pediatric OSA." *Sleep Medicine Reviews, Elsevier* (June 26, 2017). https://aomtinfo.org/wp-content/uploads/2018/05/From-oral-facial-dysfunction-to-dysmorphism-and-the-onset-of-pediatric-OSA-Guilleminault-2017.pdf.

34   Huang, Yu-Shu, Stacey Quo, J. Andrew Berkowski, and Christian Guilleminault. "Short Lingual Frenulum and Obstructive Sleep Apnea in Children." *International Journal of Pediatric Research* (2015). https://www.clinmedjournals.org/articles/ijpr/ijpr-1-003.pdf.

35   Guilleminault, Christian, Shehlanoor Huseni, and Lauren Lo. "A frequent phenotype for pediatric sleep apnoea: short lingual frenulum." *ERJ Open Research, U.S. National Library of Medicine, National Institutes of Health* (July 2016). https://www.ncbi.nlm.nih.gov/pmc/articles/PMC5034598/.

36   Garcia, Patricia, Jennifer Haile. "Notes from the Field: Lead Poisoning in an Infant Associated with a Metal Bracelet—Connecticut, 2016." *Morbidity and Mortality Weekly Report, Centers for Disease Control and Prevention* (September 1, 2017). https://www.cdc.gov/mmwr/volumes/66/wr/mm6634a6.htm.

37   Welch, Ashley. "FDA Warns About Teething Jewelry After 18-Month-Old Dies." CBS News (December 21, 2018). https://www.cbsnews.com/news/fda-warns-about-teething-jewelry-after-18-month-old-dies/.

38   American Dental Association. "Thumb-Sucking, Pacifier Use May Damage
      Children's Teeth." *Science News, Science Daily* (December 25, 2001). https://www
      .sciencedaily.com/releases/2001/12/011224083205.htm.

39   "Baby Bottle Tooth Decay." *Mouth Healthy, American Dental Association* (2020).
      https://www.mouthhealthy.org/en/az-topics/b/baby-bottle-tooth-decay.

40   Delta Dental. "2011 Children's Oral Health Survey" (2011). https://www
      .deltadentalnm.com/deltaDentalNewMexico/files/8a/8aa12164-bc79-41a9-af08
      -5d5b135ac878.pdf.

41   Burhenne, Mark. *Dr. B's Guide to Preventing & Reversing Cavities Naturally [FOR
      KIDS].* Ask the Dentist. https://guideforkids.safechkout.net.

CHAPTER 3

1   Gelb, Michael, and Howard Hindin. *Gasp: Airway Health—The Hidden Path
      to Wellness.* Scotts Valley, CA: CreateSpace Independent Publishing Platform
      (December 2016), 36.

2   "Largest Consumer Sleep Study Ever Released at CES 2017." SleepScore Labs
      (January 6, 2017). https://www.sleepscore.com/news/largest-consumer-sleep
      -study-ever-released-ces-2017/.

3   Hublin, Christer, Markku Partinen, Markku Koskenvuo, and Jaako Kaprio. "Sleep
      and Mortality: A Population-Based 22-Year Follow-Up Study." *Sleep, Oxford
      Academic* (October 2007), 1245–53. https://academic.oup.com/sleep/article/30/10
      /1245/2696836.

4   Gorvett, Zaria. "What You Can Learn from Einstein's Quirky Habits." *BBC Future*
      (June 12, 2017). https://www.bbc.com/future/article/20170612-what-you-can-learn
      -from-einsteins-quirky-habits.

5   Popova, Maria. "Thomas Edison, Power-Napper: The Great Inventor on Sleep and
      Success." *Brain Pickings* (June 1, 2012). https://www.brainpickings.org/2013/02/11
      /thomas-edison-on-sleep-and-success/.

6   Lin, Steven. *The Dental Diet.* Carlsbad, CA: Hay House Publishing (2019), refs. 12, 13.

7   Lieberman, Daniel E. *The Evolution of the Human Head.* Cambridge, MA: Belknap
      Press (2011).

8   Lin, Steven. *The Dental Diet,* chapter 3, ref. 10.

9   Nitric Oxide—The Miracle Molecule." The Nitric Oxide Society (2018). https://www
      .nitricoxidesociety.org.

10   Lieberman, Daniel E. *The Evolution of the Human Head.*

11   Martins, Diana Lopes Lacerda, Luciana Fontes Silva Cunha Lima, Vanessa
      Favero Demeda, Ana Luiza Oliveira da Silva, Angela Rosanne Santos de Oliveira,
      Flávia Melo de Oliveira, Sarah Beatriz Freire Lima, and Valéria Soraya de Farias
      Sales. "The Mouth Breathing Syndrome: Prevalence, Causes, Consequences and
      Treatment." *Journal of Surgical and Clinical Research* (2014). https://periodicos.ufrn
      .br/jscr/article/view/5560.

12 Boivin, Diane B. *Sleep and You: Sleep Better, Live Better*. Toronto, Canada: Dundurn Press (2014).

13 Ingram, David G. *Sleep Apnea in Children: A Handbook for Families*. Scotts Valley, CA: CreateSpace Independent Publishing Platform (2018).

14 Herlin, Bastien, Smaranda Leu-Semenescu, Charlotte Chaumereuil, and Isabelle Arnulf. "Evidence That Non-dreamers Do Dream: A REM Sleep Behavior Disorder Model." *Journal of Sleep Research* (August 25, 2015). https://onlinelibrary.wiley.com/doi/full/10.1111/jsr.12323.

15 National Institutes of Health. "Brain May Flush Out Toxins During Sleep" (October 17, 2013). https://www.nih.gov/news-events/news-releases/brain-may-flush-out-toxins-during-sleep.

16 Ingram, David. *Sleep Apnea in Children: A Handbook for Families*, ref 3.

17 Boivin, Diane B. *Sleep and You: Sleep Better, Live Better*.

18 Perfect, Michelle M., Kristen Archbold, James L. Goodwin, Deborah Levine-Donnerstein, and Stuart F. Quan. "Risk of Behavioral and Adaptive Functioning Difficulties in Youth with Previous and Current Sleep Disordered Breathing." *Sleep* (April 1, 2013), 517–25. https://academic.oup.com/sleep/article/36/4/517/2595960.

19 Bonuck, Karen, Katherine Freeman, Ronald D. Chervin, and Linzhi Xu. "Sleep-Disordered Breathing in a Population-Based Cohort: Behavioral Outcomes at 4 and 7 Years." *Pediatrics, U.S. National Library of Medicine, National Institutes of Health* (April 2012), 857–65. https://www.ncbi.nlm.nih.gov/pmc/articles/PMC3313633/.

20 Ibid.

21 Kurcinka, Mary Sheedy. *Sleepless in America: Is Your Child Misbehaving . . . or Missing Sleep?* New York: Harper Perennial (2007).

22 Westreich, Roi, Aya Gozlan-Talmor, Shahar Geva-Robinson, Tal Schlaeffer-Yosef, Tzachi Slutsky, Efrat Chen-Hendel, Dana Braiman, Yehonatan Sherf, Natan Arotsker, Yasmeen Abu-Fraiha, Liat Waldman-Radinsky, and Nimrod Maimon. "The Presence of Snoring as Well as Its Intensity Is Underreported by Women." *Journal of Clinical Sleep Medicine* (March 15, 2019). https://jcsm.aasm.org/doi/10.5664/jcsm.7678.

23 Cho, Jun-Gun, John R. Wheatley. "The Association of Carotid Artery Disease with Snoring and Obstructive Sleep Apnoea: Definitions, Pathogenesis and Treatment." *Australian Journal of Ultrasound in Medicine* (December 31, 2015). https://onlinelibrary.wiley.com/doi/full/10.1002/j.2205-0140.2010.tb00215.x.

24 Girardin, Jean-Louis, Ferdinand Zizi, Luther T. Clark, Clinton D. Brown, and Samy I. McFarlane. "Obstructive Sleep Apnea and Cardiovascular Disease: Role of the Metabolic Syndrome and Its Components." *Journal of Clinical Sleep Medicine, US National Library of Medicine, National Institutes of Health* (June 15, 2008), 261–72. https://www.ncbi.nlm.nih.gov/pmc/articles/PMC2546461/.

25 Ibid.

26 Lin, Steven. *The Dental Diet*. 52.

27   "Rising Prevalence of Sleep Apnea in U.S. Threatens Public Health."
     American Academy of Sleep Medicine (September 29, 2014). https://aasm.org/
     rising-prevalence-of-sleep-apnea-in-u-s-threatens-public-health/.

28   Graham, Tess. *Relief from Snoring and Sleep Apnea*. Scotts Valley, CA: CreateSpace
     Independent Publishing Platform (2014).

29   Huston, Kim. "Sleep Apnea Can Cause ADHD-like Symptoms in Kids. What Are
     They?" *Norton Children's* (August 27, 2019). https://nortonchildrens.com/news
     /sleep-apnea-in-children-can-cause-adhd-like-symptoms-what-are-they/.

30   O'Brien, Louise M., Neali H. Lucas, Barbara T. Felt, Timothy F. Hoban, Deborah
     L. Ruzicka, Ruth Jordan, Kenneth Guire, and Ronald D. Chervin. "Aggressive
     Behavior, Bullying, Snoring, and Sleepiness in Schoolchildren." *Sleep Medicine, US
     National Library of Medicine, National Institutes of Health* (August 12, 2011), 652–58.
     https://www.ncbi.nlm.nih.gov/pmc/articles/PMC3387284/.

31   Strickland, Rod. "Bad Teeth Number One Cause of Bullying." *Beyond Exceptional
     Dentistry* (January 8, 2014).

32   "Obesity." *CDC Healthy Schools, Centers for Disease Control and Prevention*
     (September 18, 2018). https://www.cdc.gov/healthyschools/obesity/index.htm.

33   Glick, Michael. *The Oral-Systemic Health Connection*. Batavia, IL: Quintessence
     Publishing (2014), 175, ref. 194.

34   Reilly, John J., Julie Armstrong, Ahmad R. Dorosty, Pauline M. Emmett, A. Ness,
     I. Rogers, Colin Steer, and Andrea Sherriff. "Early Life Risk Factors for Obesity in
     Childhood: Cohort Study." *The BMJ* (June 11, 2005). https://www.ncbi.nlm.nih.gov
     /pmc/articles/PMC558282/.

35   Kato, Ineko, Jose Groswasser, Patricia Franco, Sonia Scaillet, Igor Kelmanson,
     Hajime Togari, and Andre Kahn. "Developmental Characteristics of Apnea in
     Infants Who Succumb to Sudden Infant Death Syndrome." *American Journal of
     Respiratory and Critical Care Medicine* (October 15, 2001). https://www.atsjournals
     .org/doi/full/10.1164/ajrccm.164.8.2009001.

36   Rambaud, C., and C. Guilleminault. "Death, Nasomaxillary Complex, and Sleep in
     Young Children." *European Journal of Pediatrics* (September 2012), 1349–58. https://
     www.ncbi.nlm.nih.gov/pubmed/22492014.

37   Burger, David. "Sleep-Related Breathing Disorder Treatment Outlined in New
     Policy." *ADA News,* American Dental Association.

38   "ADA Adopts Policy on Dentistry's Role in Treating Obstructive Sleep Apnea,
     Similar Disorders." American Dental Association (October 23, 2017). https://www
     .ada.org/en/press-room/news-releases/2017-archives/october/ada-adopts-policy
     -on-dentistry-role-in-treating-obstructive-sleep-apnea.

39   Huang, Yu-Shu, Stacey Quo, J. Andrew Berkowski, and Christian Guilleminault.
     "Short Lingual Frenulum and Obstructive Sleep Apnea in Children." *International
     Journal of Pediatric Research* (2015). https://www.clinmedjournals.org/articles/ijpr
     /ijpr-1-003.pdf.

40   Guilleminualt, Christian, Shehlanoor Huseni, and Lauren Lo. "A Frequent Phenotype for Pediatric Sleep Apnoea: Short Lingual Frenulum." *ERJ Open Research, US National Library of Medicine, National Institutes of Health* (July 2016). https://www.ncbi.nlm.nih.gov/pmc/articles/PMC5034598/.

41   Guimaraes, K. C., L. F. Drager, P. R. Genta, B. F. Marcondes, and G. Lorenzi-Filho. "Effects of Oropharyngeal Exercises on Patients with Moderate Obstructive Sleep Apnea Syndrome." *American Journal of Respiratory Critical Care Medicine* (May 15, 2009), 962–66. https://www.ncbi.nlm.nih.gov/pubmed/19234106.

42   Guilleminault, C., Y. S. Huang, P. J. Monteryol, R. Sato, S. Quo, and C. H. Lin. "Critical Role of Myofascial Reeducation in Pediatric Sleep-Disordered Breathing." *Sleep Medicine* (June 2013), 518–525. https://www.sciencedirect.com/science/article/abs/pii/S1389945713000658.

43   Rambaud, C., and C. Guilleminault. "Death, Nasomaxillary Complex, and Sleep in Young Children." *European Journal of Pediatrics.*

44   Prather, Aric A., Denise Janicki-Deverts, Artica H. Hall, and Sheldon Cohen. "Behaviorally Assessed Sleep and Susceptibility to the Common Cold." *Sleep* (September 1, 2015), 1353–59. https://pubmed.ncbi.nlm.nih.gov/26118561/.

45   Gelb, Michael, and Howard Hindin. *Gasp!: Airway Health—The Hidden Path to Wellness.*

46   Guilleminualt, Christian, and Farah Akhtar. "Pediatric Sleep-Disordered Breathing: New Evidence on Its Development." *Sleep Medicine Reviews* (December 2015), 46–56. https://www.ncbi.nlm.nih.gov/pubmed/26500024.

47   Christian Guilleminualt, Shannon S. Sullivan. "Towards Restoration of Continuous Nasal Breathing as the Ultimate Treatment Goal in Pediatric Obstructive Sleep Apnea." *Enliven Archive, Stanford University Sleep Medicine Division* (September 6, 2014). http://www.enlivenarchive.org/articles/towards-restoration-of-continuous-nasal-breathing-as-the-ultimate-treatment-goal-in-pediatric-obstructive-sleep-apnea.html.

48   Beninati, W., C. D. Harris, D. L. Herold, and J. W. Shepard Jr. "The Effect of Snoring and Obstructive Sleep Apnea on the Sleep Quality of Bed Partners." *Mayo Clinic Proceedings* (October 1999), 955–58. https://pubmed.ncbi.nlm.nih.gov/10918859/.

49   Sleep Group Solutions. https://join.sleepgroupsolutions.com.

50   American Migraine Foundation. "Sleep Disorders and Headache" (April 25, 2019). https://americanmigrainefoundation.org/resource-library/sleep/.

51   "Overweight & Obesity Statistics." National Institute of Diabetes and Digestive and Kidney Diseases (August 2017). https://www.niddk.nih.gov/health-information/health-statistics/overweight-obesity.

52   Klok, M. D., S. Jakobsdottir, and M. L. Drent. "The Role of Leptin and Ghrelin in the Regulation of Food Intake and Body Weight in Humans: A Review." *Obesity Reviews* (January 2007), 21–34. https://www.ncbi.nlm.nih.gov/pubmed/17212793.

53   "What Is Obstructive Sleep Apnea?" *Opti Sleep* (January 30, 2019). https://www.optisleep.com/osa/.

54 St-Onge, Marie-Pierre, Michael A. Grandner, Devin Brown, Molly B. Conroy, Girardin Jean-Louis, Michael Coons, and Deepak L. Bhatt. "Sleep Duration and Quality: Impact on Lifestyle Behaviors and Cardiometabolic Health: A Scientific Statement from the American Heart Association." *Circulation, American Heart Association* (November 1, 2016). http://circ.ahajournals.org/content/134/18/e367.

55 Gami, Apoor S., Eric J. Olson, Win K. Shen, R. Scott Wright, Karla V. Ballman, Dave O. Hodge, Regina M. Herges, Daniel E. Howard, and Virend K. Somers. "Obstructive Sleep Apnea and the Rick of Sudden Cardiac Death: A Longitudinal Study of 10, 701 Adults." *Journal of the American College of Cardiology, ScienceDirect* (August 13, 2013), 610–16. https://www.sciencedirect.com/science/article/pii/S0735109713022511?via %3Dihub.

56 "Why Sleep Apnea Raises Your Risk of Sudden Cardiac Death." *Heart & Vascular, Cleveland Clinic* (June 30, 2017). https://health.clevelandclinic.org/why-sleep-apnea -raises-your-risk-of-sudden-cardiac-death/.

57 "Stroke Awareness Month: Untreated Sleep Apnea and Stroke." Sleepapnea.org, American Sleep Apnea Association (May 7, 2017). https://www.sleepapnea.org /untreated-sleep-apnea-and-stroke-stroke-awareness-month/.

58 Lipford, Melissa C., Kelly D. Flemming, Andrew D. Calvin, Jay Mandrekar, Robert D. Brown, Virend K. Somers, and Sean M. Caples. "Associations Between Cardioembolic Stroke and Obstructive Sleep Apnea." *Sleep, US National Library of Medicine, National Institutes of Health* (November 1, 2015), 1699–1705. https://www .mayoclinic.org/documents/mc5520-0213-pdf/doc-20079143.

59 "The Link Between a Lack of Sleep and Type 2 Diabetes." sleepfoundation.org (2020). https://www.sleepfoundation.org/articles/link-between-lack-sleep-and -type-2-diabetes.

60 Boivin, Diane B. *Sleep and You: Sleep Better, Live Better,* 98.

61 Tripathi, Aunpriya, Alexey V. Melnik, Jin Xue, Orit Poulsen, Michael J. Meehan, Gregory Humphrey, Lingjing Jiang, Gail Ackermann, Daniel McDonald, Dan Zhou, Rob Knight, Pieter C. Dorrestein, and Gabriel G. Haddad. "Intermittent Hypoxia and Hypercapnia, a Hallmark of Obstructive Sleep Apnea, Alters Gut Microbiome and Metabolome." *mSystems, American Society for Microbiology* (May–June 2018). https://aomtinfo.org/wp-content/uploads/2018/06/e00020-18.full_.pdf.

62 Gupta, Vishal. "Adult Growth Hormone Deficiency." *Indian Journal of Endocrinology and Metabolism* (September 15, 2011), 197–202. https://www.ncbi .nlm.nih.gov/pmc/articles/PMC3183535/.

63 Barassi, A., M.M. Corsi Romanelli, R. Pezzilli, C.A. Damele, L. Vaccalluzzo, G. Goi, N. Papini, G.M. Colpi, L. Massaccesi, and G.V. Melzi d'Eril. "Levels of L-Arginine and L Citrulline in Patients with Erectile Dysfunction of Different Etiology." *Andrology, US National Library of Medicine, National Institutes of Health* (March 2017), 256–61. https://www.ncbi.nlm.nih.gov/pubmed/28178400.

64 "Facts and Stats." Drowsy Driving Prevention Week, National Sleep Foundation (April 20, 2019). http://drowsydriving.org/about/facts-and-stats/.

65   Tefft, B.C. "Prevalence of Motor Vehicle Crashes Involving Drowsy Drivers, United
     States, 2009–2013." AAA Foundation for Traffic Safety (November 2014). https://
     aaafoundation.org/prevalence-motor-vehicle-crashes-involving-drowsy-drivers
     -united-states-2009-2013/.

66   Bhandari, Tamara. "Sleep, Alzheimer's Link Explained." Washington University
     School of Medicine in St. Louis (July 10, 2017). https://medicine.wustl.edu/news
     /sleep-alzheimers-link-explained/.

67   "Poor Sleep Linked to Toxic Buildup of Alzheimer's Protein, Memory Loss."
     *University of California-Berkeley, EurekAlert!, American Association for the
     Advancement of Science* (June 1, 2015). https://www.eurekalert.org/pub_releases
     /2015-06/uoc--psl052915.php.

68   Osorio, Ricardo S., Tyler Gumb, Elizabeth Pirraglia, Andrew W. Varga, Shou-en
     Lu, Jason Lim, Margaret E. Wohlleber, Emma L. Ducca, Viachaslau Koushyk,
     Lidia Glodzik, Lisa Mosconi, Indu Ayappa, David M. Rapoport, and Mony J. de
     Leon. "Sleep-Disordered Breathing Advances Cognitive Decline in the Elderly."
     Neurology, American Academy of Neurology (May 12, 2015). https://n.neurology.org
     /content/84/19/1964.abstract.

69   Effland, Lara Schuster. "Depression and Sleep Problems: How to Improve Without
     Medication." Anxiety and Depression Association of America (2018). https://adaa.
     org/learn-from-us/from-the-experts/blog-posts/depression-and-sleep-problems
     -how-improve-without.

70   Sleep Group Solutions, https://join.sleepgroupsolutions.com/about-sleep-group
     -solutions/.

71   M'saad, A, I Yangui, W. Feki, N. Abid, N. Nahloul, F. MArouen, A. Chakroun, and S.
     Kammoun. "[The Syndrome of Increased Upper Airways Resistance: What Are the
     Clinical Features and Diagnostic Procedures?]" *Revue des Maladies Respiratoires*
     (December 2015), 1002–15. https://pubmed.ncbi.nlm.nih.gov/26525135/.

72   Torborg, Liza. "Mayo Clinic Q and A: Neck Size One Risk Factor for Obstructive
     Sleep Apnea." Mayo Clinic (June 20, 2015). https://newsnetwork.mayoclinic.org/
     discussion/mayo-clinic-q-and-a-neck-size-one-risk-factor-for-obstructive-sleep
     -apnea/.

73   "National Institutes of Health Sleep Disorders Research." *National Center on Sleep
     Disorders Research, National Institutes of Health* (November 2011). Planhttps://
     www.nhlbi.nih.gov/files/docs/ncsdr/201101011NationalSleepDisordersResearch
     PlanDHHSPublication11-7820.pdf.

74   Krakow, Barry, Victor A. Ulibarri, and Natalia D. McIver. "Pharmacotherapeutic
     Failure in a Large Cohort of Patients with Insomnia Presenting to a Sleep
     Medicine Center and Laboratory: Subjective Pretest Predictions and Objective
     Diagnoses." *Mayo Clinic Proceedings, PlumX Metrics* (December 2014), 1608–20.
     https://www.mayoclinicproceedings.org/article/S0025-6196(14)00473-X/abstract.

75   Lindberg, Eva, Bryndis Benediktsdottir, Karl A. Franklin, Mathias Holm,
     Ane Johannessen, Rain jogi, Thorainn Gislason, Francisco Gomez Real, Vivi

Schlunssen, and Christer Janson. "Women with Symptoms of Sleep-Disordered Breathing Are Less Likely to Be Diagnosed and Treated for Sleep Apnea Than Men." *Sleep Medicine* (July 2017), 17–22. https://pubmed.ncbi.nlm.nih.gov/28619177/.

76   Kapsimalis, F., and M. H. Kryger. "Gender and Obstructive Sleep Apnea Syndrome, Part 1: Clinical Features." *Sleep, US National Library of Medicine, National Institutes of Health* (June 15, 2002), 412–19. https://www.ncbi.nlm.nih.gov/pubmed/12071542.

77   Gelb, Michael, and Howard Hindin. *Gasp!: Airway Health—The Hidden Path to Wellness.*

78   Saaresranta, Tarja, Ulla Anttalainen, and Olli Polo. "Sleep Disordered Breathing: Is It Different for Females?" *National Center for Biotechnology Information, U.S. National Library of Medicine, National Institutes of Health* (October 2015). https://www.ncbi.nlm.nih.gov/pmc/articles/PMC5005124/.

79   Ibid.

80   Yaggi, H. Klar, John Concato, Walter N. Kernan, Judith H. Lichtman, Lawrence M. Brass, and Vahid Mohsenin. "Obstructive Sleep Apnea as a Risk for Stroke and Death." *New England Journal of Medicine* (November 10, 2005), 2034–41. https://www.nejm.org/doi/full/10.1056/NEJMoa043104.

81   "Heavy Smokers Cut Their Lifespan by 13 Years on Average." Statistics Netherlands, Netherlands Institute of Mental Health and Addiction (September 15, 2017). https://www.cbs.nl/en-gb/news/2017/37/heavy-smokers-cut-their-lifespan-by-13-years-on-average.

82   "Sleep Apnea Raises Death Risk 46 Percent: Study." Reuters, Science News (August 17, 2009). https://www.reuters.com/article/us-sleep-death-idUSTRE57H0CP20090818.

CHAPTER 4

1   Harker, LeeAnne, and Dacher Keltner. "Expressions of Positive Emotion in Women's College Yearbook Pictures and Their Relationship to Personality and Life Outcomes Across Adulthood." University of California, Berkeley (2000). http://local.psy.miami.edu/faculty/dmessinger/c_c/rsrcs/rdgs/emot/keltner_harker.yearbooksmiles.jpsp.pdf.

2   Riggio, Ronald E. "There's Magic in Your Smile." *Psychology Today* (June 25, 2012). https://www.psychologytoday.com/us/blog/cutting-edge-leadership/201206/there-s-magic-in-your-smile.

3   Savitz, Erric. "The Untapped Power of Smiling." *Forbes* (March 22, 2011). https://www.forbes.com/sites/ericsavitz/2011/03/22/the-untapped-power-of-smiling/.

4   First Things First. "Brain Development" (2021). https://www.firstthingsfirst.org/early-childhood-matters/brain-development/.

5   Scheffel, Debora Lopes Salles, Fabiano Jeremias, Camila Maria Bullio Fragelli, Lourdes Aparecida Martins dos Santos-Pinto, Josimeri Hebling, and Osmir Batista de Oliveira, Jr. "Esthetic Dental Anomalies as Motive for Bullying in

Schoolchildren." *European Journal of Dentistry* (January–March 2014). https://www.ncbi.nlm.nih.gov/pmc/articles/PMC4054024/.

6   "Poor Oral Health Can Mean Missed School, Lower Grades." Herman Ostrow School of Dentistry of USC, USC University of Southern California (August 10, 2012). https://dentistry.usc.edu/2012/08/10/poor-oral-health-can-mean-missed -school-lower-grades/.

7   "Why Students Drop Out of School: A Review of 25 Years of Research." California Dropout Research Project (October 2008). https://www.hws.edu/about/pdfs/school _dropouts.pdf.

8   "New Dove Research Finds Beauty Pressures Up, and Women and Girls Calling for Change." Dove, Cision, PR Newswire (June 21, 2016). https://www.prnewswire.com /news-releases/new-dove-research-finds-beauty-pressures-up-and-women-and -girls-calling-for-change-583743391.html.

9   Sprankles, Julie. "Research Says These Are the 5 Physical Features Men Are Attracted To Most." *Your Tango* (2017). https://www.yourtango.com/2017307563/5 -physical-features-men-find-most-attractive.

10   American Academy of Cosmetic Dentistry. "Can a New Smile Make You Appear More Successful and Intelligent?" (2021). https://aacd.com/cmsproxy/236/files /Can%20a%20new%20smile%20make%20you%20appear%20more%20 intelligent.pdf.

11   "Study Shows That One-Third of American Adults Are Unhappy with Their Smile." American Association of Orthodontists, Cision, PR Newswire (November 14, 2012), https://www.prnewswire.com/news-releases/study-shows-that-one-third-of -american-adults-are-unhappy-with-their-smile-179281261.html.

12   Shahani-Denning. "Physical Attractiveness Bias in Hiring: What Is Beautiful Is Good." Hofstra University (Spring 2003). https://www.hofstra.edu/pdf/orsp _shahani-denning_spring03.pdf.

13   "Study Shows That One-Third of American Adults Are Unhappy with Their Smile." American Association of Orthodontists, Cision, PR Newswire.

CHAPTER 5

1   Glick, Michael. *The Oral-Systemic Health Connection*. Batavia, IL: Quintessence Publishing (2014), 175.

2   Miller, W.D. "The Human Mouth as a Focus of Infection." *Dental Cosmos* (September 1891), 689–706. http://www-personal.umich.edu/~pfa/denthist/articles /Miller1891.html.

3   Kashyap, Vanita, Neha Sikka, and Reena Verma. *Periodontal and Systemic Interrelationship*. Sunnyvale, CA: Lambert Academic Publishing (2016).

4   Pallasch, Thomas J., and Michael J. Wahl. "The Focal Infection Theory: Appraisal and Reappraisal." *Journal of the California Dental Association* (March 2000), 194. https://www.cda.org/Portals/0/journal/journal_032000.pdf.

5   Glick, Michael. *The Oral-Systemic Health Connection*.

6   Ibid., 1, ref. 3.

7   Benzian, Habib, Marion Bergman, Lois K. Cohen, Martin Hobdell, and Judith Mackay. "The UN High-Level Meeting on Prevention and Control of Non-communicable Diseases and Its Significance for Oral Health Worldwide." *Wiley Online Library* (May 3, 2012). https://onlinelibrary.wiley.com/doi/abs/10.1111/j.1752-7325.2012.00334.x.

8   Glick, Michael. *The Oral-Systemic Health Connection*, 65.

9   MacKenzie, Debora. "We May Finally Know What Causes Alzheimer's—and How to Stop It." *Health* (January 23, 2019). https://www.newscientist.com/article/2191814-we-may-finally-know-what-causes-alzheimers-and-how-to-stop-it/.

10 Ibid.

11 Glick, Michael. *The Oral-Systemic Health Connection*, 78, ref 47.

12 Chen, Chang-Kai, Yung-Tsan Wu, and Yu Chao Chang. "Association Between Chronic Periodontitis and the Risk of Alzheimer's Disease: A Retrospective, Population-Based, Matched-Cohort Study." *Alzheimer's Research & Therapy* (2017). https://www.ncbi.nlm.nih.gov/pmc/articles/PMC5547465/.

13 Moraschini, Vttorio, José de Albuquerque Calasans-Maia, Mônica Diuana Calasans-Maia. "Association Between Asthma and Periodontal Disease: A Systematic Review and Meta-Analysis." *Journal of Periodontology, American Academy of Periodontology* (February 23, 2018). https://aap.onlinelibrary.wiley.com/doi/10.1902/jop.2017.170363.

14 Glick, Michael. *The Oral-Systemic Health Connection*.

15 World Health Organization. "Human Papillomavirus (HPV) and Cervical Cancer." (November 11, 2020). https://www.who.int/news-room/fact-sheets/detail/human-papillomavirus-(hpv)-and-cervical-cancer.

16 Ibid., 96.

17 Dr. Oz. "The Surprising Link Between Gum Disease and Cancer" (2021). https://www.doctoroz.com/blog/surprising-link-between-gum-disease-and-cancer.

18 *Science Daily*. "Link Found Between Periodontal Disease and Pancreatic Cancer" (January 17, 2007). https://www.sciencedaily.com/releases/2007/01/070116205547.htm.

19 Tezal, M., M. A. Sullivan, A. Hyland, J. R. Marshall, D. Stoler, M. E. Reid, T. R. Loree, N. R. Rigual, M. Merzianu, L. Hauck, C. Lillis, J. Wactawski-Wende, and F. A. Scannapieco. "Chronic Periodontitis and the Incidence of Head and Neck Squamous Cell Carcinoma." *Cancer Epidemiology, Biomarkers & Prevention* (September 2009), 2406–12. https://www.ncbi.nlm.nih.gov/pubmed/19745222.

20 Glick, Michael. *The Oral-Systemic Health Connection: A Guide to Patient Care*.

21 Koren, Omry, Ayme Spor, Jenny Felin, Frida Fak, Jesse Stombaugh, Valentina Tremaroli, Carl Johan Behre, Rob Knight, Bjorn Fagerberg, Ruth E. Ley, and Fredrik Backhed. "Human Oral, Gut, and Plaque Microbiota in Patients with Atherosclerosis." *Proceedings of the National Academy of Sciences of the United*

States of America (March 15, 2011) 4592–98. https://pubmed.ncbi.nlm.nih.gov
/20937873/.

22   Choi, H. M., K. Han, Y. G. Park, and J. B. Park. "Associations Among Oral Hygiene
Behavior and Hypertension Prevalence and Control: The 2008 to 2010 Korea
National Health and Nutrition Examination Survey." *Journal of Periodontology,
US National Library of Medicine, National Institutes of Health* (July 2015) 866–73.
https://www.ncbi.nlm.nih.gov/pubmed/25741579.

23   Matthews, Debora. "Possible Link Between Periodontal Disease and Coronary
Heart Disease." *Evidence-Based Dentistry* (March 25, 2008). https://www.nature
.com/articles/6400560.

24   Kshirsagar, Abhijit, Kevin L. Moss, John R. Elter, James D. Beck, Steve Offenbacher,
and Ronald J. Falk. "Periodontal Disease Is Associated with Renal Insufficiency
in the Atherosclerosis Risk in Communities (ARIC) Study." *American Journal of
Kidney Diseases* (April 2005), 650–57. https://pubmed.ncbi.nlm.nih.gov/15806467/.

25   Ricardo, Ana C., Ambarish Athavale, Jinsong Chen, Hemanth Hampole, Daniel
Garside, Philip Marucha, and James P. Lash. "Periodontal Disease, Chronic Kidney
Disease and Mortality: Results from the Third National Health and Nutrition
Examination Survey." *BMC Nephrology* (2015). https://www.ncbi.nlm.nih.gov/pmc
/articles/PMC4492086/.

26   Cedars-Sinai. "Diabetes and Gum (Periodontal) Disease" (2021). https://www
.cedars-sinai.org/health-library/diseases-and-conditions/d/diabetes-and-gum
-periodontal-disease.html.

27   Diabetes.co.uk. "Nearly Half of Adults with Type 2 Diabetes Worldwide Are Unaware
of Their Condition" (November 14, 2016). https://www.diabetes.co.uk/news/2016/
nov/nearly-half-of-adults-with-type-2-diabetes-worldwide-are-unaware-of-their
-condition-94586467.html.

28   Doyle, Kathryn. "Chronic Gum Disease Tied to Risk of Erectile Dysfunction."
Reuters (November 25, 2016). https://www.reuters.com/article/us-health
-periodontitis-erectile-dysfunc/chronic-gum-disease-tied-to-risk-of-erectile
-dysfunction-idUSKBN13K1UP.

29   OSH News Network. "Periodontal Therapy Comparable to 30% Drop in LDL
Cholesterol" (January 14, 2016). http://oshnewsnetwork.com/2016/01/14/
periodontal-therapy-comparable-to-30-drop-in-ldl-cholesterol/.

30   Yu, Hui-Chieh, Tsung-Po Chen, and Yu-Chao Chang. "Inflammatory Bowel
Disease as a Risk Factor for Periodontitis Under Taiwanese National Health
Insurance Research Database." *Journal of Dental Sciences* (September 2018) 242–47.
https://www.sciencedirect.com/science/article/pii/S1991790218303192.

31   Vavricka, S. R., C. N. Manser, S. Hediger, M. Vögelin, M. Scharl, L. Biedermann, S.
Rogler, F. Seibold, R. Sanderink, T. Attin, A. Schoepfer, M. Fried, G. Rogler, and P.
Frei. "Periodontitis and Gingivitis in Inflammatory Bowel Disease: A Case-Control
Study." *Inflammatory Bowel Diseases* (December 2013), 2768–77. https://www.ncbi
.nlm.nih.gov/pubmed/24216685.

32 OSH News Network. "P.G. Promotes Progression of Fatty Liver Disease." (March 3, 2016). http://oshnewsnetwork.com/2016/03/03/pg-fatty-liver-disease/.

33 Benahmaed, Asma Gasmi, Amin Gasmi, Alexandru Dosa, Salvatore Chirumbolo, Pavan Kumar Mujawdiya, Jan Aaseth, Maryam Dadar, and Geir Bjorklund. "Association Between the Gut and Oral Microbiome with Obesity." *Anaerobe* (August 2021). https://www.sciencedirect.com/science/article/abs/pii/S1075996420301049.

34 Esfahanian, Vahid, Mehrnaz Sadighi Shamami, and Mehrnoosh Sadighi Shamami. "Relationship Between Osteoporosis and Periodontal Disease: Review of the Literature." *Journal of Dentistry of Tehran University of Medical Sciences* (Autumn 2012), 256–64. https://www.ncbi.nlm.nih.gov/pmc/articles/PMC3536461/.

35 Glick, Michael. *The Oral-Systemic Health Connection*, 187.

36 American Academy of Periodontology. "New Study Links Periodontitis and COVID-19 Complications" (February 3, 2021). https://www.perio.org/periodontitis_COVID-19_complications.

37 Darrah, Erika. "Gum Disease Linked to Rheumatoid Arthritis." *Johns Hopkins Rheumatology* (January 23, 2017). https://www.hopkinsrheumatology.org/2017/01/gum-disease-linked-to-rheumatoid-arthritis/.

38 Christensen, Thor. "How Oral Health May Affect Your Heart, Brain and Risk of Death." *American Heart Association News* (March 19, 2021). https://www.heart.org/en/news/2021/03/19/how-oral-health-may-affect-your-heart-brain-and-risk-of-death.

39 Bengtsson, Viveca Wallin, Gosta Rutger Persson, Johan Sanmartin Berglund, and Stefan Renvert. "Periodontitis Related to Cardiovascular Events and Mortality: A Long-Time Longitudinal Study." *Clinical Oral Investigations* (January 28, 2021). https://link.springer.com/article/10.1007/s00784-020-03739-x.

40 Roizen, Michael F. *RealAge: Are You as Young as You Can Be?* New York: William Morrow (1999). https://www.amazon.com/RealAge-Are-You-Young-Can/dp/0060191341.

41 Paganini Hill, Annila, Stuart C. White, and Kathryn A. Atchison. "Dental Health Behaviors, Dentition, and Mortality in the Elderly: The Leisure World Cohort Study." *Journal of Aging Research* (June 15, 2011). https://pubmed.ncbi.nlm.nih.gov/21748004/.

42 Rosengard, Heather C. "Oral Manifestations of Systemic Diseases." *Medscape* (June 29, 2018). https://emedicine.medscape.com/article/1081029-overview.

43 Eckel, Ashley, Dale Lee, Gail Deutsch, Anthony Maxin, and Dolphine Oda. "Oral Manifestations as the First Presenting Sign of Crohn's Disease in a Pediatric Patient." *Journal of Clinical and Experimental Dentistry* (July 2017), 934–38. https://www.ncbi.nlm.nih.gov/pmc/articles/PMC5549595/.

44 Environmental Working Group. "Body Burden: The Pollution in Newborns" (July 14, 2005). https://www.ewg.org/research/body-burden-pollution-newborns.

45 Ibid.

46 Curatola, Gerald P., and Diane Reverand. *The Mouth-Body Connection: The 28-Day Program to Create a Healthy Mouth, Reduce Inflammation and Prevent Disease Throughout the Body*. New York: Center Street (2017).

47 Grandjean, Philippe. "Developmental Fluoride Neurotoxicty: An Updated Review." *Environmental Health* (December 19, 2019). https://www.ncbi.nlm.nih.gov/pmc /articles/PMC6923889/.

48 Amaechi, Bennet T., Parveez Ahamed AbdulAzees, Dina Ossama Alshareif, Marina Adel Shehata, Patricia Paula de Carvalho Sampaio Lima, Azadeh Abdollahi, Parisa Samadi Kalkhorani, and Veronica Evans. "Comparative Efficacy of a Hydroxyapatite and a Fluoride Toothpaste for Prevention and Remineralization of Dental Caries in Children." *BDJ Open* (December 9, 2019). https://www.nature.com /articles/s41405-019-0026-8.

49 Pajor, Kamil, Lukasz Pajchel, and Joanna Kolmas. "Hydroxyapatite and Fluorapatite in Conservative Dentistry and Oral Implantology—A Review." *Materials* (August 22, 2019). https://www.mdpi.com/1996-1944/12/17/2683.

50 Schroth, R. J., R. Rabbani, G. Loewen, and M. E. Moffat. "Vitamin D and Dental Caries in Children." *Journal of Dental Research* (February 2016), 173–79. https:// pubmed.ncbi.nlm.nih.gov/26553883/.

51 Garcia, M. Nathalia, Charles F. Hildebolt, S. Douglas Miley, Debra A. Dixon, Rex A. Couture, Catherine L. Anderson Spearie, Eric M. Langenwalter, William D. Shannon, Elena Deych, Cheryl Mueller, and Roberto Civitelli. "One-Year Effects of Vitamin D and Calcium Supplementation on Chronic Periodontitis." *Journal of Periodontology* (January 2011), 25–32. https://pubmed.ncbi.nlm.nih.gov/20809866/.

52 The Royal Children's Hospital Melbourne. "Essential Oil Poisoning" (July 2021). https://www.rch.org.au/clinicalguide/guideline_index/Essential_Oil_Poisoning/.

53 National Center for Complementary and Integrative Health. "Peppermint Oil" (August 22, 2021).

54 Doran, Anna Louise, and Joanna Verran. "A Clinical Study on the Effect of the Prebiotic Inulin in the Control of Malodour." *Microbial Ecology in Health and Disease* (February 14, 2007). https://www.tandfonline.com/doi/full/10.1080 /08910600701521279.

55 Soderling, Eva, and Kaisu Pienihakkinen. "Effects of Xylitol and Erythritol Consumption on Mutans Streptococci and the Oral Microbiota: A Systematic Review." *Acta Odontologica Scandinavica* (July 7, 2020). https://www.tandfonline .com/doi/full/10.1080/00016357.2020.1788721.

56 De Cock, Peter, Kauko Makinen, Eino Honkala, Mare Saag, Elke Kennepohl, and Alex Eapen. "Erythritol Is More Effective Than Xylitol and Sorbitol in Managing Oral Health Endpoints." *International Journal of Dentistry* (August 21, 2016). https:// www.ncbi.nlm.nih.gov/pmc/articles/PMC5011233/.

57 Boronow, Katherine E., Julia Green Brody, Laurel A. Schaider, Graham F. Peaslee, Laurie Havas, and Barbara A. Cohn. "Serum Concentrations of PFASs and Exposure-Related Behaviors in African American and Non-Hispanic White

Women." *Journal of Exposure Science & Environmental Epidemiology* (2019), 206–17. https://www.nature.com/articles/s41370-018-0109-y.

CONCLUSION

1 Chamie, Joseph. "America's Single-Parent Families." *The Hill* (March 19, 2021). https://thehill.com/opinion/finance/543941-americas-single-parent-families.

2 Bureau of Labor Statistics, U.S. Department of Labor. "Employment Characteristics of Families—2020" (April 21, 2021). https://www.bls.gov/news.release/pdf/famee .pdf.

3 Otto, Mary. "How Can a Child Die of Toothache in the US?" *The Guardian* (June 13, 2017). https://www.theguardian.com/inequality/2017/jun/13/healthcare-gap-how -can-a-child-die-of-toothache-in-the-us.

4 Raphael, John. "Simple Toothache Turns Deadly, Killing a Young Father from Sacramento." *Nature World News* (February 1, 2017). https://www.natureworldnews .com/articles/35339/20170201/simple-toothache-turns-deadly-killing-young -father-sacramento.htm.

5 "What Is Dental Anxiety and Phobia?" Colgate, Reviewed by the Faculty of Columbia University College of Dental Medicine (September 18, 2013). https:// www.colgate.com/en-us/oral-health/basics/dental-visits/what-is-dental-anxiety -and-phobia.

6 Normando, David. "How Much Is It Worth a Smile?" *Dental Press Journal of Orthodontics* (May–June 2015), 11–12. https://www.ncbi.nlm.nih.gov/pmc/articles /PMC4520132/.

7 "Achieve a Healthy Smile with Delta Dental." Delta Dental Insurance (2018). https://www.rahndentalgroup.com/dental-care-plus-dental-insurance.html.

INDEX

ABOUT THE AUTHOR

 Both artist and scientist, Dr. Kami Hoss is a Renaissance man: a classical composer and musician, a renowned orthodontist, and an entrepreneur on a mission to reinvent the oral health care paradigm in America. Armed with a masters in craniofacial biology from USC, a doctorate in dental surgery from UCLA, and a post-doctorate in orthodontics and dentofacial orthopedics, he co-founded The Super Dentists—one of the country's leading multi-specialty dental practices—with his business partner and wife, Dr. Nazli Keri.

Over the last twenty-five years, The Super Dentists has been at the forefront of oral care innovation, utilizing the latest tools and techniques to provide patients with the safest, fastest, and most effective treatments ever. They even cut down the time it takes for braces to straighten teeth, thanks to Dr. Hoss's invention of Acceledontics™—a breakthrough system that realigns teeth in a fraction of the time required by traditional braces. More recently, Dr. Hoss has launched SuperMouth, a line of revolutionary oral care products with custom tools and solutions for every age and stage of development.

Besides providing his patients with exceptional care, Dr. Hoss offers community programs, seminars, and workshops all over the country. His speaking engagements focus on oral health and its impact on pregnancy, sleep, disease, and even emotional well-being, giving people the tools and information to dramatically improve their lives.

But opening patients' eyes is just one piece of the puzzle. Frustrated with the poor level of education at the current dental assisting schools in California, Dr. Hoss founded Howard Healthcare Academy® in San Diego. In partnership with The Super Dentists, Howard Healthcare Academy trains students in an unprecedented hands-on method, as well as providing educational programs for doctors, focusing on integrating oral care into whole body health care.

Dr. Hoss is a member of the American Association of Orthodontists, the American Dental Association, the California Dental Association, the Forbes Business Council, and the Newsweek Expert Forum. He sits on the board of counselors at UCLA School of Dentistry and is a highly sought-after expert who has been featured on NBC, ABC, FOX, CBS, and NPR affiliates nationally, as well as in hundreds of newspapers, magazines, and other media outlets. *If Your Mouth Could Talk* is his first book.